Justine Mondry

Lateral organization of cell-matrix proteins on the nanoscale

Chapter 7

Lateral organisation of cell matrix proteins on the nanoscale

Justine Mondry

Lateral organization of cell-matrix proteins on the nanoscale

Südwestdeutscher Verlag für Hochschulschriften

Impressum / Imprint
Bibliografische Information der Deutschen Nationalbibliothek: Die Deutsche Nationalbibliothek verzeichnet diese Publikation in der Deutschen Nationalbibliografie; detaillierte bibliografische Daten sind im Internet über http://dnb.d-nb.de abrufbar.
Alle in diesem Buch genannten Marken und Produktnamen unterliegen warenzeichen-, marken- oder patentrechtlichem Schutz bzw. sind Warenzeichen oder eingetragene Warenzeichen der jeweiligen Inhaber. Die Wiedergabe von Marken, Produktnamen, Gebrauchsnamen, Handelsnamen, Warenbezeichnungen u.s.w. in diesem Werk berechtigt auch ohne besondere Kennzeichnung nicht zu der Annahme, dass solche Namen im Sinne der Warenzeichen- und Markenschutzgesetzgebung als frei zu betrachten wären und daher von jedermann benutzt werden dürften.

Bibliographic information published by the Deutsche Nationalbibliothek: The Deutsche Nationalbibliothek lists this publication in the Deutsche Nationalbibliografie; detailed bibliographic data are available in the Internet at http://dnb.d-nb.de.
Any brand names and product names mentioned in this book are subject to trademark, brand or patent protection and are trademarks or registered trademarks of their respective holders. The use of brand names, product names, common names, trade names, product descriptions etc. even without a particular marking in this works is in no way to be construed to mean that such names may be regarded as unrestricted in respect of trademark and brand protection legislation and could thus be used by anyone.

Coverbild / Cover image: www.ingimage.com

Verlag / Publisher:
Südwestdeutscher Verlag für Hochschulschriften
ist ein Imprint der / is a trademark of
OmniScriptum GmbH & Co. KG
Heinrich-Böcking-Str. 6-8, 66121 Saarbrücken, Deutschland / Germany
Email: info@svh-verlag.de

Herstellung: siehe letzte Seite /
Printed at: see last page
ISBN: 978-3-8381-3673-8

Zugl. / Approved by: Dortmund, TU, Diss., 2014

Copyright © 2014 OmniScriptum GmbH & Co. KG
Alle Rechte vorbehalten. / All rights reserved. Saarbrücken 2014

Contents

Abstract   vii

I   Introduction   1

1 Biological Background   3
- 1.1 Cell-matrix adhesion 4
- 1.2 The integrin receptor 4
  - 1.2.1 Integrin: structure and activation 5
- 1.3 Nascent adhesion formation 10
  - 1.3.1 Initial binding proteins 10
- 1.4 Focal adhesion maturation 13
  - 1.4.1 Focal complexes 15
  - 1.4.2 Focal adhesion sites 15
- 1.5 Adhesion regulation by GTPases 17
- 1.6 Focal adhesion disassembly 18
- 1.7 Composition and organization of focal adhesions 19
  - 1.7.1 Nano-scale organization 20

2 Objectives   21

3 Theoretical Aspects of Microscopy   23
- 3.1 Light 23
  - 3.1.1 Fluorescence 23
  - 3.1.2 Monochromatic light 24
- 3.2 Diffraction barrier of light microscopy 25
  - 3.2.1 Overcoming the diffraction limit in theory 25
- 3.3 Super-resolution microscopy 26
  - 3.3.1 Photoactivated localization microscopy 27
  - 3.3.2 Total internal reflection fluorescence 28
  - 3.3.3 PALM procedure 29

II Methods 31

4 General biological methods 33
4.1 DNA handling 33
4.1.1 Cloning strategy 33
4.1.2 Bacterial DNA expression 33
4.1.3 Site-directed mutagenesis 35
4.2 Cell culture 35
4.2.1 General cell line handling 35
4.2.2 Transient transfection 36
4.3 Experimental cell preparation 36
4.3.1 Preparation of cell dishes for microscopy 36
4.3.2 Fixation and staining of REF52 cells 37
4.3.3 Inhibition of mechanical force 37
4.3.4 Induction of Integrin clustering 37

5 Data acquisition & analysis 39
5.1 Photoactivated localization microscopy 39
5.1.1 Protocol of data acquisition 40
5.1.2 Data analysis 41
5.2 Single particle tracking 44
5.2.1 Data aquisition for $\beta 3$-integrin tracking 45
5.2.2 Data acquisition of spot activation 45
5.2.3 Data analysis 46

III Results 47

6 Super-resolution imaging of adhesion sites 49
6.1 Spatial organization 49
6.1.1 Degree distribution of simulations 50
6.1.2 Degree distribution of adhesion proteins 52
6.1.3 Influence of the expression level 56
6.1.4 Nano-polarity of single focal adhesions 58
6.2 Spatial organization in single focal adhesions 60
6.3 Protein nano-clusters in adhesion sites 61
6.3.1 Molecular density degree in adhesion sites 62
6.3.2 Cluster analysis 66
6.4 Speculations concerning protein accumulations 70

7 Single particle tracking in living cells — 73
- 7.1 Mobility changes of $\beta 3$-integrin upon force inhibition — 73
  - 7.1.1 $\beta 3$-integrin dynamics influenced by force inhibition — 74
- 7.2 Density alterations upon force inhibition — 78
  - 7.2.1 Dynamic distribution of $\beta 3$-integrin in adhesion sites — 79
- 7.3 Induction of the high-affinity state of integrin — 83
- 7.4 Formation of new focal adhesion sites — 86

8 Dense domains as potential signaling centers — 91
- 8.1 Spreading efficiency of adhesion proteins — 91
- 8.2 FAK and Paxillin recruitment to distinct areas — 93

IV Discussion and future perspectives — 97

9 — 99
- 9.1 Nano-organization in fixed cells — 99
  - 9.1.1 Analysis of adhesion protein organization — 99
  - 9.1.2 Dual-color PALM — 101
  - 9.1.3 Protein localization influenced by force inhibition — 101
  - 9.1.4 Protein localization outside adhesion sites — 102
- 9.2 Nano-organization in living cells — 102
  - 9.2.1 Influence of actomyosin contractility — 103
  - 9.2.2 Diffusion-driven density formation — 105
- 9.3 Directed protein recruitment — 106

Acknowledgments — 109

References — 111

List of Figures — 128

List of abbreviations — 131

Zusammenfassung

Die Verankerung einer Zelle wird durch eine Ansammlung von multifunktionellen Proteinen gewährleistet, welche Fokale Adhäsionen genannt wird. In dieser Arbeit wurde die räumliche Anordnung einiger dieser Proteine untersucht, indem die super-auflösende Mikroskopietechnik PALM angewendet wurde. Dabei wurde gezeigt, dass alle Adhäsionsproteine definierte Bereiche unterschiedlicher Dichte innerhalb von Fokalen Adhäsionen ausbilden. Proteinakkumulationen mit sehr hoher Dichte können mehrere Dutzend Moleküle enthalten und einen Durchmesser von über hundert Nanometern erreichen. Eine übergeordnete Struktur, geschweige denn eine Polarität konnte für solche Proteinansammlungen jedoch nicht nachgewiesen werden.

Des Weiteren wurden die Proteinakkumulationen auf ihre Entstehung und ihre zeitlichen Veränderungen am Beispiel des Adhäsionsrezeptors $\beta 3$-Integrin untersucht. Dabei wurde gezeigt, dass eine molekulare Umverteilung, welche auch zu einer Neuorientierung der Dichte führt, durch Kraftinhibierung hervorgerufen werden kann. Ein kompletter Abbau der $\beta 3$-Integrinakkumulationen wird durch Kraftinhibierung dagegen nicht verursacht. Dementsprechend ist davon auszugegehen, dass Aktomyosinkontraktionen eine modulierende, jedoch keine induzierende Funktion hat. Stattdessen liegt es nahe, dass die Grundsteinlegung einer $\beta 3$-Integrinakkumulation bereits im frühen Entwicklungsstadium erfolgt. Der zeitliche Verlauf des Entstehungsprozesses Fokaler Adhäsionen wurde anhand von $\beta 3$-Integrin verfolgt und beobachtet, dass sich bereits anfänglich gebildete Cluster mit fortschreitendem Adhäsionswachstum zu größeren Ansammlungen ausbreiten können.

Außerdem wurde gezeigt, dass das Signalprotein FAK bevorzugt zu flächenlimitierten Gebieten in einer Fokalen Adhäsion rekrutiert wird, was für eine erhöhte biologische Aktivität sprechen würde.

Abstract

Cellular anchors are large accumulations of a multitude of multi-functional proteins and are known as focal adhesion sites. In this work, the spatial organization of a subset of cell-matrix proteins was analyzed, using the super-resolution microscopy technique PALM. It was demonstrated, that all cell-matrix proteins form distinct areas of varying densities inside single focal adhesions. Highly dense protein accumulations can contain up to several tens of molecules and can span a diameter of more than hundred nanometers. However, no recognizable structure or polarity could be observed for such large protein accumulations.

In order to study the temporal alterations and formation of such highly dense protein accumulations, the dynamic behavior of the adhesion receptor β3-integrin was analyzed. It was shown, that force inhibition can induce structural rearrangements, also leading to the redistribution of the density inside adhesion sites. However, force inhibition did not cause the complete disassembly of dense β3-integrin domains. Therefore, it is suggested that force can modulate the dense areas, but is not the initial inducer. Instead, it seems that an initial formation of dense domains occurs already in the very beginning of focal adhesion development, followed by a gradual immobilization of β3-integrin and thus leading to the formation of new adhesion sites. It was observed that the seed for some dense domains can be planted already in early maturation stages with the potential to increase its size upon adhesion expansion.

Dense domains could even have a particular signaling function, as it was shown that the signaling protein FAK is primarily recruited to delimited areas inside focal adhesions, which could represent dense domains.

Part I
Introduction

Chapter 1

Biological Background

For the development of cellular systems, individual cells must be able to interact with each other to form a functional unit. This implies adhesion to the surrounding cells and the environment as a fundamental feature. Adhesion plays a major role in the survival of most cell types, as detachment induces a shutdown of many survival transduction signals and is usually followed by apoptosis.

Directly related to adhesion and of comparable importance is cell migration. Only mobile cells can insert themselves into an organized multicellular environment which is essential for initial embryonic development. Many other processes can be carried out only by the interplay of motility and adhesion like epidermal wound healing in vertebrates for example. Just minutes after an injury, mobile platelets are recruited from the bloodstream to the injured tissue. Here, they adhere and form a plug which temporary stops the bleeding. Furthermore, they secrete inflammatory factors which supports proliferation and migration of other bloodstream cells, like leukocytes and neutrophils. Days after the injury, fibroblasts and endothelial cells from the surrounding tissue start to migrate chemotactically towards the wound. The invading fibroblasts initiate the healing process by providing extracellular matrix components which attract cellular adhesion. Epithelial cells immobilize and close the wound permanently [6, 129].

Thus, a functional immune system correlates with a balanced adhesion system and can be easily perturbed by anomalies in the adhesion system [62, 5, 48, 125]. Metastatic spreading of cancer is directly connected to cell motility and often accompanied by mutations of the proteins involved in adhesion [118, 26, 162]. The significance of adhesion - not only in diseases but also in the evolution of life in general - has led to a growing scientific interest to understand the basic mechanisms of the controlled directional attachment of cells.

1.1 Cell-matrix adhesion

Principally, adhesion is enabled by cellular affection for the surrounding environment which is called the extracellular matrix (ECM). The components of the ECM are produced and secreted by exocytosis in multicellular structures and aggregate to a gel-like structure [54]. Fibrous proteins and proteoglycans are the main components of the ECM, but the particular composition is unique for each tissue. According to the distinct composition, the function of the ECM varies from providing mechanical strength and stabilization to the activation of signaling cascades, because it can also embed cytokines and growth factors [120]. These ECM inhomogeneities are crucial for the formation of functional tissue, as it induces polarity within a cell. Polarity represents the first step of directed migration and can be separated into chemotaxis [156] which is induced by signaling molecules, and haptotaxis [27] caused by an ECM density gradient. As soon as the ECM composition alters towards conditions favoring adhesion, motile cells embed in terms of tissue formation. Therefore, the cellular reaction can encompass stable anchorage to the ECM, as well as changes in morphology or migration and proliferation, depending on the conditions provided by the ECM.

The most important receptor family linking the cell and the ECM is represented by integrins [69]. Integrins recognize multiple ECM components as ligands and serve as the first sensors of extracellular mechanical signals [61]. Furthermore, integrins induce the recruitment of cytoplasmic proteins which build large plaques, called focal adhesion sites [145].

1.2 The integrin receptor

The first link between the ECM and the cell is the transmembrane receptor integrin [138] which consists of an α and a β subunit. So far, out of 18 α- and 8 β-subunits, 24 α/β distinct combinations are known to form functional receptor heterodimers [91, 68]. Some of these subunits are exclusively expressed in specific cell types (such as $\beta 2$ and $\beta 7$ in leukocytes), but a high binding affinity for ECM is shared by all of them. A frequently recognized ligand is the peptide sequence arginyl-glycyl-aspartic acid (RGD) which is part of many different ECM proteins. However, the integrin binding affinity highly depends on the α-β combination, as RGD is usually not the only interaction site. For instance, it was shown that the RGD-containing ECM-protein Vitronectin is recognized with high affinity by $\beta 3$-heterodimers, whereas $\beta 1$-heterodimers bind with significant lower affinity [56, 86]. The opposite is true for Fibronectin, another RGD-motif-protein [102, 168].

1.2. THE INTEGRIN RECEPTOR

A unique characteristic of integrin receptors is their bidirectional activation mechanism. Besides the classical ligand-induced outside-in activation mechanism, integrin receptor activation can also occur the other way around (inside-out) by the assembly of cytosolic binding proteins at the integrin-tail. Among the various intracellular binding partners (directly and indirectly) are several kinases which enable signal transduction, because integrin receptors contain no intrinsic enzymatic activity [152]. To gain a better understanding of the complex activation steps, a closer look at the molecular structure of the integrin receptor is useful.

1.2.1 Integrin: structure and activation

Since the beginning of the century, many crystal structures [164, 165, 163], NMR studies [146, 25, 75] and electron microscopy observations [4, 137] contributed largely to understand the domain structure of integrin-heterodimers, and its role in the activation mechanism. Generally, all integrin subunits consist of a large extracellular head domain, a single-spanning transmembrane region and a short cytoplasmic tail which contains binding sites for cytosolic proteins [87].

The ectodomain of the β-subunit

The extracellular section of the β-subunit of integrin can be described as a globular head standing on a rather flexible stalk. This stalk consists of a membrane-proximal β-tail domain, 4 EGF-repeats, a hybridized (plexin-semaphorin-integrin) PSI-domain and a hybrid-domain, followed by the head region. The decisive domain for ligand binding and receptor activation is the *inserted domain* (I-domain) which is inserted in the hybrid domain, which is in turn inserted into the PSI domain. Together these three domains form the globular head of the β-subunit. The I-domain consists of 7 α-helices and 6 centered β-sheets [164]. Serine residues within the connecting loop of $\alpha 1$ and $\beta 1$, and an aspartate of the $\beta 4\alpha 5$-loop coordinate physiologically a Mg^{2+} ion in the center of the I-domain. This cation binding site is named MIDAS (metal-ion-dependent adhesive site) and is flanked by an ADMIDAS (adjacent to MIDAS) and a SyMBS (synergistic metal ion-binding dependent adhesion site) which both bind preferentially Ca^{2+} [31]. The ADMIDAS appears to function as a negative regulator of integrin activation, since Ca^{2+} ions have an inhibiting effect [113]. At the same time, the Ca^{2+} occupied SyMBS synergizes activation [16]. Upon ligand binding, the Mg^{2+} in the MIDAS alters its coordination towards the acidic ligand residue (often glutamic acid) and induces steric changes in the organization of the secondary

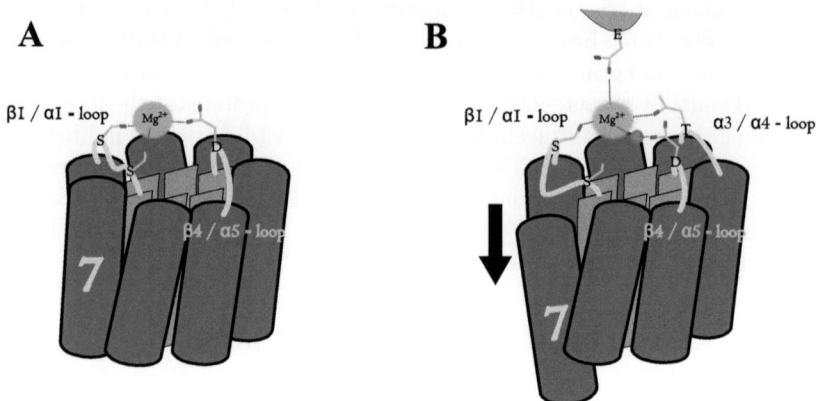

Figure 1.1: I-domain of the integrin receptor.
The Mg^{2+} ion in the MIDAS changes its coordination upon ligand binding. The transition from the unbound (A) to the ligand-bound (B) state induces a mechanical downward movement of the α7-helix.

structure of the I-domain, as illustrated in Figure 1.1. The $\alpha1\beta1$-loop alters its position but both serine residues remain coordinated, whereas the aspartate of the $\beta4\alpha5$-loop is only indirectly connected to Mg^{2+} via a water molecule. Instead, a threonine of the $\alpha3\alpha4$-loop is directly coordinated. The ligand-induced rearrangement of the loops involved in cation coordination causes a shift of the $\beta6\alpha7$-loop (not shown in Figure 1.1) leading to an axial downwards movement of the $\alpha7$-helix by 7 Å. In order to compensate the $\alpha7$-helix shift the following hybrid-domain of the receptor-subunit changes its position by a swing-out mechanism. This induces the separation from the loosely connected integrin α-subunit and a solidification of the following stalk-domains [91, 58, 173].

The ectodomain of the α-subunit

The N-terminus of the α-subunit consists of a large β-propeller head, formed by seven blade-like arranged β-sheets. Half of the α-subunits also contain an I-domain including the ligand-coordinating MIDAS but lack the regulatory ADMIDAS and SyMBS. The I-domain of the α-subunit is inserted between β-sheet 2 and 3 of the β-propeller, and regulates ligand binding independently from the β-subunit. However, also the β-subunit requires a Mg^{2+}-recoordination within the MIDAS to induce the mandatory swing-out of the hybrid-domain. This is achieved by a conserved glutamic acid residue

1.2. THE INTEGRIN RECEPTOR

in the loop between α7-helix and β-sheet 3, which is pushed into the MIDAS of the β-subunit, where it functions as a pseudo-ligand. The β-propeller is followed by the rather stalk region consisting of a thigh-domain, which is separated by a flexible linker from two static calf-domains [4, 173].

Ligand binding

The ligand binding interface is formed by the conserved interaction of β-subunit's I-domain with the upper part of α-subunit's β-propeller. This interaction is supported by an additional segment in the head domain of the β-subunit [105, 31].

In a heterodimer without an α-I-domain (e.g. $\alpha v \beta 3$), the ligand binds primarily to the interface of both subunits, becoming coordinated with the metal ion in the MIDAS. RGD-ligands for example, interact via their aspartic acid-residue with the MIDAS and bind with the arginine residue to an aspartate-residue within the β-propeller. In receptors containing an α-I-domain (e.g. $\alpha 5 \beta 1$), the ligand binds only to the β-propeller of the α-subunit, which results in a distinct binding specificity to the ECM components Laminin and Collagen, but rather not to RGD-peptides. The β-subunit contributes also to ligand binding specificity, because it possesses a *specificity determining loop*, which significantly affects the affinity for distinct ligands [137, 91, 152].

Conformational change upon activation

It was already mentioned that ligand binding induces a mechanical displacement of the C-terminal α7-helix within the I-domain, which causes a swing-out of the β-subunit hybrid-domain. This movement effects the overall conformation of the integrin receptor.

There are at least 3 distinguishable conformational states of integrin, which are directly correlated with the activation state, as shown in Figure 1.2. In the inactive, or low-affinity-state, the integrin-heterodimer remains in a closed conformation. Thereby, the head domain is bent to the membrane proximal leg domain, enabled by a 120° flexion of the linkers between the thigh and the calf domains in the α-subunit and the first and second EGF-module in the β-subunit. It is worth noting that a Ca^{2+} ion is coordinated to the genu of the α-subunit. In this V-shape position, the ligand binding site is unpropitiously oriented, which results in a lower probability for ligand binding. In the bent conformation, both leg-regions are in close proximity, as are the transmembrane α-helices, which are tightly packed and undergo canonical glycine-glycine interactions within the membrane. The C-termini of both subunits form a conserved salt-bridge in the membrane proximal

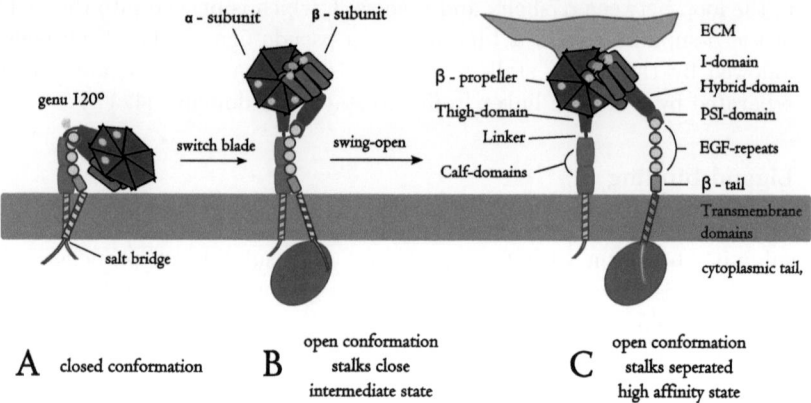

Figure 1.2: Integrin activation.
The inactive state is characterized by a closed conformation (A). Thereby, both ectodomains are in a membrane proximal, bent position. The extracellular rearrangement of the Mg^{2+} ion in the I-domain, as well as the intracellular disruption of the salt bridge between both tail domains result in a switch-blade opening of the receptor. This is not always associated with a tight ligand binding and can create an intermediate state, in which both leg regions are in close proximity (B). A permanent ligand binding induces the swing-open of the hybrid-domain in the β-subunit, which is always followed by the separation of both leg-domains. This conformation represents the extended high-affinity, ligand bound state (C).

region, including Arg^{995} of the alpha-subunit and Asp^{747} of the beta-subunit, which prevents binding of cytosolic proteins [115, 163]. However, leg and tail interactions are weak and can be broken easily by inside-out activation [92].

The active, or high-affinity state is characterized by a completely extended receptor with explicitly separated stalk-regions. The conformational switch of the head-domains occurs at the genu, which functions as a pivot and follows a switch-blade mechanism. The switch is initially induced by the swing-open movement of the hybrid-domain which is generally connected with ligand binding. Apart from the outside-in activation, the switch can be induced also by intracellular disruption of the weak salt-bridge connection by intracellular binding proteins. Consequently, the cytosolic integrin tails separate, leading to a general connection loss of transmembrane-domains and sometimes even of the stalk-regions. In this case, the mechanical switch takes place because the high-affinity conformation, including the hybrid-domain-

1.2. THE INTEGRIN RECEPTOR

swing, is energetically favored. If the stalks do not separate, an intermediate state can be observed of an already extended receptor but with the flexible β-subunit-stalk in loose proximity to the α-subunit-stalk, resulting in insufficient ligand binding an thus an non-extended hybrid-domain. Usually, the state of high affinity is reached as soon as tight ligand binding occurs [91].

Inside-out-activation via Talin

Relevant interactions have only been observed with the short cytosolic tail of the β-subunit, whereas the tail of the α-subunit seems to have no function in focal adhesion formation. The tail of the β-subunit contains the conserved motifs NPxY and NxxY, which represent binding sites for phosphotyrosine-binding (PTB)-domains. In addition, a less conserved membrane-proximal HDRK motif also functions as potential binding site [85]. Considering that the high-affinity state is energetically preferred when the integrin tail domains are separated, an inside-out-activation mechanism is logical. Only Talin has been shown to induce this tail separation by directly interacting with the cytosolic tail of the β-subunit and causing the salt-bridge breakage [151].

Talin consists of an N-terminal FERM-domain (which contains the integrin binding site), an unstructured linker region including a Calpain cleavage site and a long rod-domain, which contains 14 helix-bundles with the potential to unfold and extend the protein up to 60 nm [174]. The atypical extended FERM-domain in the globular head can be separated in four subunits: F0, F1, F2 and F3. The F3 subunit contains a PTB-like domain and has high affinity to the conserved NxxY site on the β-C-tail of integrin, proximal to the membrane. As a result, Talin competes with the salt-bridge binding and eventually disrupts it. Head binding of Talin leads to a separation of the receptor-subunits and induces the high-affinity-state of integrin [153].

It is surprising that only Talin is able to induce inside-out activation, as there are several PTB-containing proteins with an affinity towards integrin, but do not activate the receptor *in vivo*. This unique ability of Talin could be related to its additional hydrophobic pocket in the F3-region, which interacts with two phenylalanine residues (723 and 730 in the β3-subunit) in the integrin-C-terminus. This supporting interaction could enhance the binding stability and additionally have a catalytic effect on the salt-bridge disruption. It was shown that also F0 and F1 are required for integrin activation although the mechanism is not known [19, 24].

Furthermore, Talin needs supporting proteins for a successful integrin activation. Kindlin1, 2 and 3, which consist mainly of a Talin-like FERM domain, seem to have a potential effect on inside-out-signaling, however this has not been well studied yet. The role of the Kindlin family proteins could be to displace potential inhibitory proteins from the β-subunit-tail, which eases Talin access, or a direct binding to NxxY of the β-subunit to enhance the exposure of Talin's binding motif [92, 74, 24]. Additionally, recruitment of cytosolic Talin to the membrane is a critical regulation factor, which is mediated through several proteins.

1.3 Nascent adhesion formation

Integrin receptors can be understood as cellular anchoring tools with the additional function of sensing the proximal environment. Usually, the formation of new anchors (focal adhesion sites) occurs in the lamellipodia of a polar cell during migration [2, 71]. Typical for these broad cellular protrusions is a dense meshwork of fibrillar actin (F-actin), whose characteristic branching is induced by the Arp2/3 complex [6, 111, 59]. The continuous polymerization of F-actin in the cellular front induces the extension of the lamelipodia by pushing the membrane in the direction of cell motility. This is accompanied by a backwards movement of the F-actin meshwork, known as retrograde flow [141].

The protruding membrane contains a high level of integrin receptors in the high-affinity state [79, 81], resulting in ligand recognition in the ECM. The ligand binding causes integrin receptor immobilization, which is the initial step of cellular anchorage. The immobilized integrins form micro-clusters, which entail the independent recruitment of intracellular cell-matrix adhesion proteins, creating a link to the actin meshwork [51]. An accumulation of first binding proteins at the cytoplasmic tail of the integrin receptor is known as a nascent adhesion [144] and represents the precursor of a potential focal adhesion site. Among these first binding partners are Focal Adhesion Kinase (FAK), Src and Paxillin, as well as Talin and Kindlins. Cytoskeletal proteins like Vinculin and α-actinin have been reported in nascent adhesions too.

1.3.1 Initial binding proteins

More than 50 different proteins are known to assemble in focal contacts [168]. In order to keep it simple, only the ones relevant for this work will be discussed here and are illustrated in Figure 1.3. The assembly of cytosolic cell-matrix adhesion proteins at the cytosolic part of the integrin receptor is generally strongly supported by creating a membrane environment that is enriched in phosphoinositol-4,5-bisphosphate (PIP2) [154]. The formation of a nascent adhesion is illustrated in Figure 1.4.

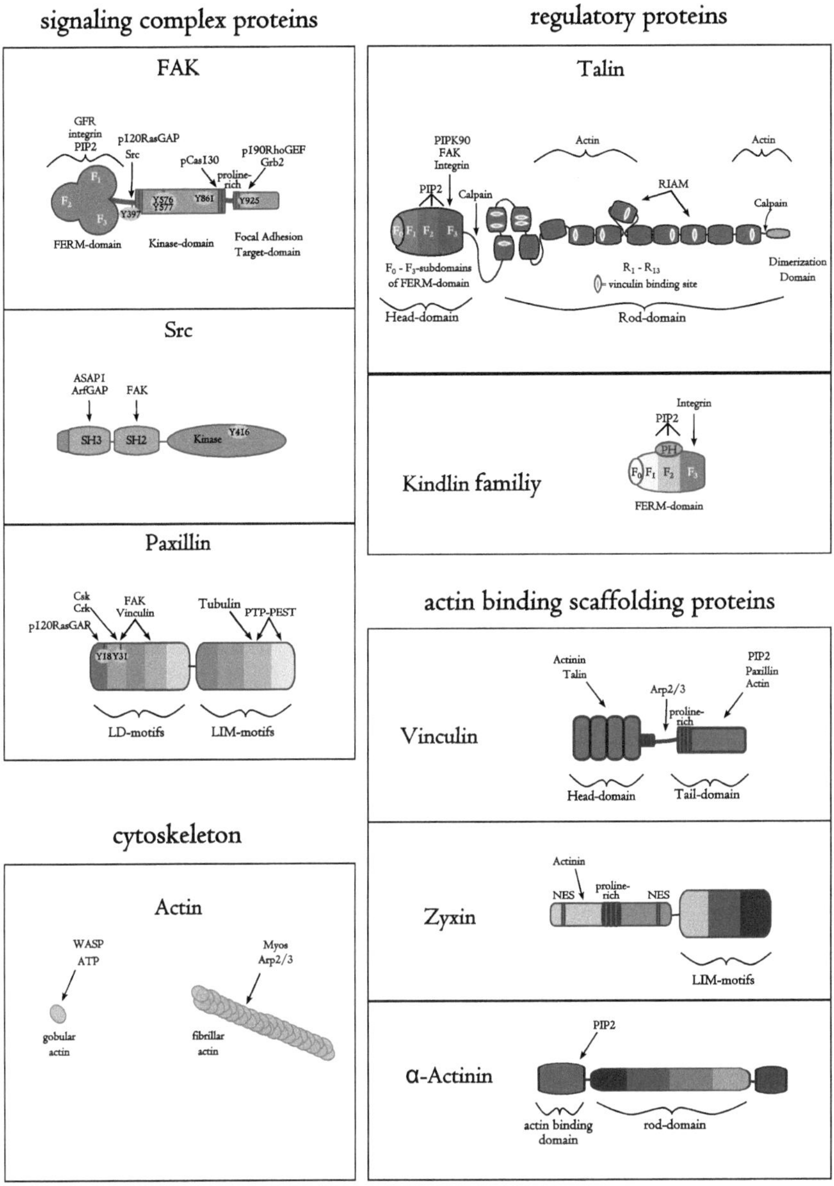

1.3. NASCENT ADHESION FORMATION

Figure 1.3: Focal adhesion proteins.
This Figure presents all cell-matrix proteins that are relevant in this work. The illustrated binding sites and phosphorylation sites are explained in detail in this chapter.

Talin and Kindlins

PIP2 recruits the Rap1-GTP-interacting adaptor molecule (RIAM) to the membrane together with Talin, as RIAM binds to the rod-domain of Talin [128, 24]. However, a potential receptor activator like Talin must be well regulated. This is ensured by an intramolecular auto-inhibition state, in which the ninth helix-bundle of the rod-domain interacts with the F3-subunit of the FERM domain and therefore masks the integrin binding site (cf. section 1.2.1). When located at a PIP2-rich membrane, the negative charge attracts the head domain and repulses the rod domain, which activates Talin and ensures an advantageous positioning for integrin binding [135, 148]. The released rod-domain reveals several binding sites for F-actin, which mark Talin as the direct connection between the receptor and the cytoskeleton. A connection to the branched meshwork of F-actin potentially functions as a supporting scaffold for the nascent adhesion [107, 145]. The Kindlins, as the second activators also show interactions with PIP2, as well as with PIP3 [166], but the recruitment mechanism is still elusive.

FAK and Src

The tyrosine-kinase FAK is also present in nascent adhesions and was shown to bind to the β-subunit of integrin with its PTB-like domain [121, 32]. Like Talin, FAK rests in the cytosol in an auto-inhibited state, in which the N-terminal FERM domain interacts with the central kinase domain, while the FAT (focal adhesion target) domain is exposed to the surrounding cytosol. Due to the high affinity of the FAT-domain towards Talin and Paxillin [108, 52], FAK can be positioned in close proximity to the membrane. The exact recruitment mechanism is still not known, and might be unrelated to the FAT-domain interaction with Talin/Paxillin. Recent studies suggest that FAK could function as the very first integrin-interaction partner - even upstream of Talin [82, 83] - which must involve another recruitment mechanism. Inactive FAK was shown to interact directly with the Arp2/3-complex and could represent an alternative mechanism for membrane localization in the lamellipodia [126].

A membrane-proximal localization of FAK is further supported by an interaction of the FERM domain with PIP2, which triggers a conformational change [101, 23]. As a result, the kinase domain becomes accessible for either *cis* or *trans* auto-phosphorylation of Tyr^{397}. $pTyr^{397}$ is a highly specific binding site for the SH2-domain of Src tyrosine kinase and recruits Src to nascent adhesions [52]. Src-binding induces a cascade of phosphorylation events within FAK, including Tyr^{576} and Tyr^{577} in the activation loop [101].

The FAK-Src-complex is the center of kinetic activity in an adhesion site as integrin receptors themselves lack an intrinsic kinase domain [69].

Paxillin

The scaffold protein Paxillin belongs also to the first interacting proteins in nascent adhesions [45, 127] and is furthermore a substrate of the FAK-Src-complex [142, 14]. Paxillin had numerous binding partners because it contains several LIM (zinc finger) domains and LD (leucine and aspartate) motifs [21] as well as SH2 and SH3 (Src homology 2 and 3) binding sites. All these domains are potential protein-binding sites. Therefore, it is not exactly clear which protein is targeted by Paxillin in the nascent complex. However, the existence of these multiple binding sites underscores Paxillin's role as an important protein organizer in focal adhesions [42].

Vinculin and α-actinin

The adhesion proteins Vinculin and α-Actinin were also mentioned as part of a nascent adhesion, which is surprising as these proteins are known as cytoskeletal scaffold proteins [13, 70]. Since both proteins contain binding sites for PIP2, their membrane-recruitment and therefore their early appearance in nascent adhesions is explicable [168, 34, 145]. Further, Vinculin seems to interact directly with the Arp2/3 complex [43] resulting in its localization to the branched F-actin meshwork. In addition, both proteins have distinct binding sites for F-actin. Still, the question remains whether these proteins interact already in a functional way with other cell-matrix proteins in a nascent adhesion.

1.4 Focal adhesion maturation

Many nascent complexes form in the lamellipodia, but most of them disassemble (turn-over) within seconds when they enter the lamella, which represents the border to the cell body [8]. The transition of nascent adhesions to focal complexes and then to focal adhesions is rather vague and floating. Not least, because many of the key proteins of a focal adhesion are already present in nascent adhesions and focal complexes. The fate of a nascent adhesion or a focal complex depends a lot on the functionality of the recruited proteins, as well as on the linkage to the actin filaments and is not yet completely understood. Still, some characteristics are specific for each state, which allows to draw the line between the different states. Figure 1.5 illustrates schematically the transition into a maturing focal adhesion site.

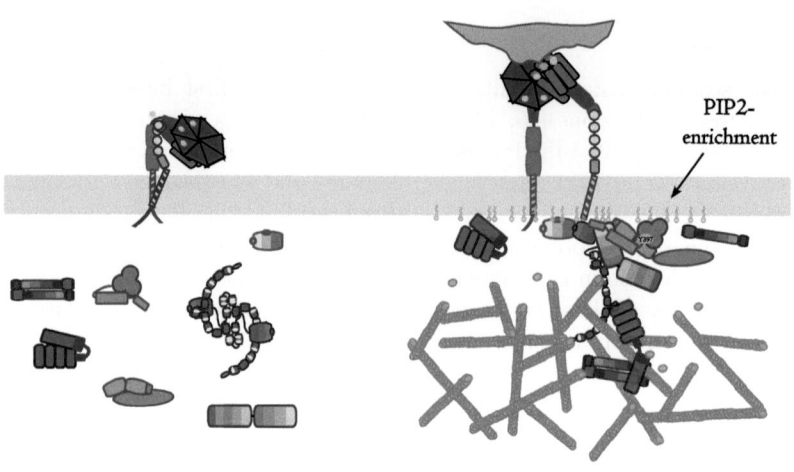

Figure 1.4: Nascent adhesion formation.
All proteins are illustrated in detail in Figure 1.3.
In the inactive state of the transmembrane integrin receptor, the intracellular cell-matrix adhesion proteins are usually not membrane-localized. In the cytosol, Talin, FAK, Src and Vinculin exist in an auto-inhibited closed conformation (left).
In the lamellipodia, the branched actin meshwork interacts potentially with Vinculin, α-actinin and FAK, which could represent a way to recruit these proteins. Integrin activation induces the recruitment of cell-matrix proteins, highly supported by PIP2 enrichment of the membrane (right). Talin, FAK, Vinculin and α-actinin were shown to interact with PIP2. Kindlin, Talin and FAK can interact with integrin directly. The illustration represents a possible composition, derived from known protein-protein-interactions. However, the exact molecular interactions and composition in this state are not known yet. However, the release of the auto-inhibition of FAK marks the first step of the formation of the signaling complex. FAK, Src and Paxillin interact tightly and regulate the signaling transduction. It is most probable that this complex is functional already in nascent adhesions.

1.4. FOCAL ADHESION MATURATION

1.4.1 Focal complexes

The protein composition of a nascent adhesion overlaps to a large extent with the one of a focal complex. What differs is the cellular localization, the size and the application of force. Whereas the F-actin meshwork of the lamellipodia polymerizes very rapidly and independently, the lamella-entrance ensures a connection to the force-controlled cytoskeleton. Only few of the F-actin strands formed in the lamellipodia will be gripped by the actomyosin-network and potentially rescue nascent adhesions from turn-over, whereas all other branched F-actin strands depolymerize again, together with most of the nascent adhesions.

The actomyosin traction force, applied to the remaining nascent adhesions in the early lamella induces the extension of Talin's rod-domain [39], which could represent the transition point to a focal complex. In turn, several binding sites for the head-domain of Vinculin become exposed in Talin's rod, leading to its recruitment. Vinculin switches from a closed to an open conformation, combined with the release of its tail-domain. The exposed C-terminus has binding sites for F-actin as well as for Paxillin and assists in scaffold and stability maintenance. The traction force in the early lamella is significantly higher than in all other parts of the cell [17], and imposes a high pulling force on the focal complexes. Depending on the integrin subunits, as well as on the ligand, a single integrin receptor has an adhesion strength of 5-30 pN [88, 112, 80]. Consequently the focal complex needs to recruit more cytosolic cell-matrix proteins, which enhance also integrin clustering [36]. As a result, the focal complex starts to grow, marking the initial phase of focal adhesion maturation. Focal complexes have a lifetime of a few minutes depending on the migration ability of the cell as they exist only in the junction of lamellipodia and early lamella. Focal complexes mature either into focal adhesions or disassemble as a consequence of failing to build a stabilizing environment [60, 145, 9].

1.4.2 Focal adhesion sites

When the surviving focal complex passes the lamellipodia border, the initial traction force decreases and induces Zyxin recruitment [17]. The mechano-sensing protein Zyxin functions as a reinforcement factor for actin stress fiber bundling and as a recruitment factor for α-actinin. α-actinin forms antiparallel dimers and thereby connects actin filaments. Both proteins support the building of actin bundles, which in turn causes the enlargement of the anchorage by recruiting more adhesion proteins. This process is generally known as focal adhesion maturation and mainly defined by a growing size.

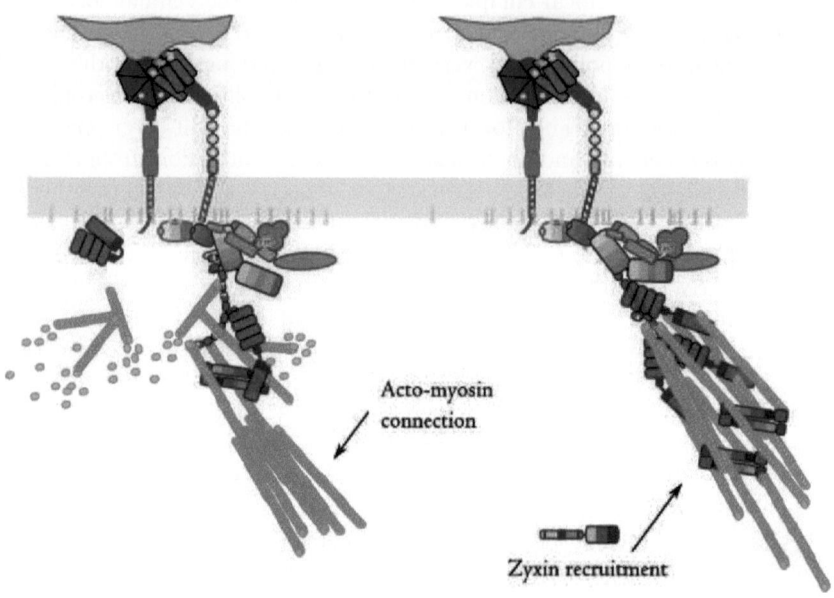

Figure 1.5: Maturation process of focal adhesions.
All proteins are illustrated in detail in Figure 1.3.
A nascent adhesion transforms into a focal complex by resisting mechanical force, which is applied in the early lamella. This force is provided by the myosin-driven actin cytoskeleton, which connects with F-actin strands of the branched actin meshwork. Non-connected actin filaments will depolymerize at the lamellipodia-lamella border (left). The actomyosin connection stretches the rod-domain of Talin, which exposes Vinculin binding sites. The recruitment of Vinculin and other cell-matrix proteins enlarges the focal complex and marks the maturation process into focal adhesions. The strong initial traction force diminishes and Zyxin is recruited (right), which defines the transition of a focal complex into a maturing focal adhesion.
This illustration represents one of several possible molecular compositions.

Due to the centripetal movement of the actin bundles, also the focal adhesion maturation orients in the centripetal direction and forms oval shapes, which differ in size, depending on the tension applied. Focal adhesion sites can persist for hours as predominantly steady structures and allow the cell body to slide over them to enable migration. Focal adhesions have fixed positions relative to the ECM but undergo continuous fluctuation in size and shape [169]. These natural changes are accompanied by alterations in F-actin bundle size, force and traction, but do not automatically lead to turn-over of focal contact sites.

1.5 Adhesion regulation by GTPases

The small GTPases Cdc42, Rac1 and RhoA are the major controlling enzymes of the cytoskeletal organization and thereby influence migration and adhesion of cells [65]. Cell migration is supported by Cdc42-induced polarity and Rac1-mediated formation of lammelipodiae, promoting furthermore the formation of nascent adhesions and focal complexes. An enhancement of the Rac1 level triggers remodeling of the actin cytoskeleton, towards less stable actin bundles resulting in a weakening of focal adhesion sites [152].

On the other hand, cell adhesion is sustained by a strong actin cytoskeleton, mediated by RhoA and the formation of solid focal adhesion sites [140]. The RhoA downstream effector ROCK (Rho-associated protein kinase) enhances the stability of focal adhesions by phosphorylating the myosin light chain and synergistically inhibiting myosin light chain phosphatase. As a result, myosin activity is catalyzed, leading to enhanced actomyosin traction force. This supports the formation of mature focal adhesions, which can lead to the creation of large plaques. The GTPase-typical activity-switch is induced by GEF's ("on") and GAP's ("off"), which are partly regulated by the FAK-Src complex of focal contact sites [42].

Strong evidence suggests that high phosphorylation rates and the resulting signaling appears already in nascent adhesions [35]. Thereby, mainly the Cdc42 and Rac1 pathway become activated in order to stimulate migration. FAK-Src induces phosphorylation of p190RhoGAP, ASAP1(ARF-GTPase-activating protein 1) and GRAF (GTPase regulator associated with FAK) and turns these proteins into active RhoA inhibitors. Furthermore, the complex p130Cas-Csk functions as a Rac1GEF. This complex is modulated by pTyr^{861}FAK-Src-complex interaction. Also Paxillin recognizes pTyr^{861}FAK with its LD-motifs and is consequently phosphorylated at Tyr^{31} and Tyr^{118} upon binding. This supports formation of the PaxillinGITPIX-complex, a downstream signal for Rac activity [170]. Cdc42 activity is enhanced by

FAK-Src mediated phosphorylation of N-WASP (neuronal WiskottAldrich syndrome protein), which is a substrate of Cdc42 and enhances Actin branching in the lamellipodia [160].

The FAK-Src-complex can support RhoA activity as well, by activating p190RhoGEF and PDZRhoGEF. The phosphatase PTP-PEST binds to the LIM-motif of Paxillin and thereby inhibits signaling via PaxillinGITPIX [86]. PTP-PEST also inactivates p190RhoGAP via dephosphorylation, which enhances RhoA activity [72].

It is still not known how the balance between Rac and Rho is exactly maintained [61], however, it was shown that the constitution of integrin receptors inside a focal adhesion plays a key role [99, 167]. Furthermore, focal adhesion signaling influences also the signal transduction of Ras. A direct link to proliferation enhancement is provided by capturing p120RasGAP, which hinders its inhibiting function for active Ras. GRB2 (Growth factor receptor-bound protein) is known to support Ras activation and the ERK2 (extracellular signal-regulated kinase-2) cascade and can be bound by pTyr^{925}FAK-Src [122]. Apparently, GRB2 binding excludes Paxillin binding and is connected with Dynamin2 recruitment [50], which might support turn-over of focal adhesions.

1.6 Focal adhesion disassembly

In a migrating cell, detachment is as important as adhesion, which includes the concerted work of several processes. It can be distinguished between a turn-over of focal contacts in the leading edge, a reduction of focal adhesion in the steady state by changes in the traction force, and a complete turn-over of large focal contact sites, which are positioned in the cell rear after the cell body has moved by.

Generally, focal adhesion dynamics are mainly controlled by the organization of the cytoskeleton, which includes alterations in the balance of RhoA, Rac1 and Cdc42. The level of PIP2 in the membrane is another decisive factor for adhesion protein recruitment and their proper membrane-coordination. PIP2 synthesis is provided by PIPKIγ, which in turn is a regulating factor for Talin [44, 104]. FAK-Src-mediated phosphorylation of PIPKIγ enhances the binding affinity to F3 in the head-domain of Talin, which competes directly with integrin-binding [124, 84] and consequently destabilizes the adhesion complex. The FAK-Src-complex phosphorylates furthermore Tyr^{12} in α-actinin, which weakens the actin binding affinity. As a result, F-actin becomes destabilized due to reduced cross-linking [70] which also diminishes focal adhesions. Disassembly of the FAK-Src-complex can be induced by

1.7. COMPOSITION AND ORGANIZATION

ERK2-mediated phosphorylation of Ser^{910} in FAK, causing a significant reduction of focal adhesions. The Paxillin-associated phosphatase PTP-PEST is able to dephosphorylate FAK at $pTyr^{397}$ [171, 172], representing the initial phosphotyrosine in FAK which is mandatory for activity [98]. The phosphorylation pattern of Paxillin, as well as its LD4-motif, seems to play an essential role in focal adhesion turn-over but little is unraveled yet [150].

However, for large focal adhesion plaques more effort is required in order to remove the tight anchorage of the cell. As already mentioned, the large GTPase Dynamin2 can be recruited to focal adhesions by the FERM-domain of FAK and can be bridged by GRB2 in order to become phosphorylated by Src [149]. Active Dynamin2 is responsible for endocytosis and may cause the partial internalization of focal adhesions [22]. Dynamin2 is moreover associated with microtubuli which also promote focal adhesion disassembly [78, 50]. Microtubuli seem to target mainly focal adhesion sites at the late cell body and the trailing rear and involve clathrin-dependent endocytosis, however with unknown mechanism [49, 46]. At the cellular rear, enhanced levels of Ca^{2+} recruit the protease Calpain to focal adhesions, where it finds cleavage sites between the head and the rod domain of Talin [53], as well as between the kinase and FAT domain of FAK [29]. Vinculin and Paxillin have also been reported to be Calpain targets [89, 38].

1.7 Composition and organization of focal adhesions

Focal adhesion sites contain a large number of different proteins. The proteins described in this chapter represent only some of the well studied proteins, but are not necessarily more important than other non-mentioned proteins in terms of their signaling or scaffolding function.

Such a complex structure can only work as a functional unit if all of its components work in concert. Many cell-matrix proteins contain binding domains that are potential targets for several proteins (cf. Figure 1.3). However, no simultaneous binding is possible, which allows several different molecular organizations. Mechanical force is essential in focal adhesion development and has been demonstrated to function as a modulator of focal adhesion size and shape [159]. However, it was not revealed whether the general protein composition alters upon traction force. It was shown that in nascent adhesions and focal complexes, $\alpha 5\beta 1$-integrin is predominantly active, whereas mature focal adhesions are mainly anchored by $\alpha v\beta 3$-integrin [168, 77]. The receptor and the ECM composition of a focal contact strongly modulates the signaling events of Rho and Rac [99, 167]. Since integrin receptors have no intrinsic kinase domain, the regulation of small GTPases must be provided by recruited kinases and phosphatases. Selective recruitment to the cytosolic integrin-tail only can generate significant alterations of the signal transduction. This selective recruitment could change the relative composition of all cell-matrix proteins, not just the signaling proteins. However, this question has not been addressed in detail, so far.

1.7.1 Nano-scale organization

Many cell-matrix proteins are well studied, with respect to their domain structure and their binding partners. Still, the molecular localization inside focal adhesions is mostly unknown. The intrinsic organization of adhesion sites has always been an active field of research and is still not completely unraveled. However, recent advancements in microscopy techniques pave the way for big steps forward.

Recent super-resolution studies revealed a distinct axial position for several cell-matrix proteins in adhesion plaques [76, 33]. This suggests that focal adhesions assemble in a highly organized fashion in order to build a functional unit. Also *in vivo* experiments could obtain a fair nano-resolution in adhesion sites [131] despite their highly dynamic nature, which supports a general nano-organization of adhesion proteins within a focal adhesion complex. Additional super-resolution observations of several adhesion proteins exposed a relative spatial localization to each other in distinct areas [132]. However, the overall pattern of molecular organization has not been studied yet.

Chapter 2
Objectives

Focal adhesion sites have a crucial function in cellular anchorage due to their static positioning. Furthermore, they can persist for hours without showing major changes in size and shape. This suggests a highly organized structure in order to maintain their adhesive function, as high fluctuation rates could negatively influence their stability and weaken the anchorage.

So far, little was done to propose a spatial nano-organization of focal adhesion proteins in lateral directions [117, 133], which will be discussed in the first part of this work. The focal adhesion receptor β3-integrin, the adhesion activating proteins Kindlin1, Kindlin2 and Talin, the adhesion signaling proteins FAK, Src and Paxillin, as well as the cytoskeletal proteins Vinculin, α-actinin, Zyxin and β-actin will be investigated in terms of their general structural features inside focal adhesion sites. For this purpose, the super-resolution technique *photo-activated localization microscopy* (PALM) will be used to visualize the nano-scale localization of individual proteins. These localizations will be further analyzed by a localization proximity analysis, which is inspired by degree distribution analysis [40].

In the second part of this work, the nano-organization of focal adhesions will be studied in living cells. For this purpose, the dynamic behavior of the receptor β3-integrin will be observed, using *single particle tracking* (SPT) analysis. The lateral rearrangements of β3-integrin will be visualized under conditions of focal adhesion assembly and disassembly, using localization proximity analysis. Furthermore, the directed recruitment of the signaling proteins FAK and Paxillin to focal adhesions will be studied in order to find potential signaling centers.

Chapter 3

Theoretical Aspects of Microscopy

The localization and the dynamics of proteins in a living system can be observed using fluorescence microscopy. However, this is severely limited by the spatial resolution in the lateral and axial dimensions, which is dictated by the diffraction limit. For that reason, advanced microscopy methods have been developed to circumvent the diffraction limit.

In this chapter, the basics of fluorescence microscopy are described, followed by a more detailed explanation of the super-resolution technique used in this work.

3.1 Light

Light is defined as electromagnetic radiation, which is limited to the wavelengths visible to the human eye of about 400-800 nanometers. Light consists of photons, which are defined energy bundles and show the physical features of waves as well as the ones of particles. This is known as the wave-particle-duality, and allows the application of the wave or the particle model, depending on the experiment.

3.1.1 Fluorescence

The phenomenon of fluorescence appears when a fluorophore emits light after the previous absorption of a photon. It can be understood via the particle model and illustrated by a Jablonski diagram in Figure 3.1 [7].

The electrons of a fluorophore are usually in the ground state (S_0), consisting of several vibrational energy levels with the ability to enter higher energy levels (S_1, S_2, ...) by absorbing energy. This excitation energy can be provided by a photon with a distinct wavelength, matching the energy needed for entering a vibrational level of a higher energy state. After excitation, the electron loses some energy within picoseconds and drops back via vibrational relaxation to the first vibrational state of S_1. This energy level has a lifetime of a few nanoseconds until the electron finally drops back into the ground state, emitting a photon. The emitted light is always shifted to a longer wavelength (Stokes shift)[94, 116], because of the energy loss induced by vibrational relaxation. In rare cases, an electron can flip its spin, which transfers the electron to a triplet state (intersystem crossing). The relaxation to the ground state is spin forbidden, resulting in an extended lifetime of seconds to hours, visible as phosphorescence.

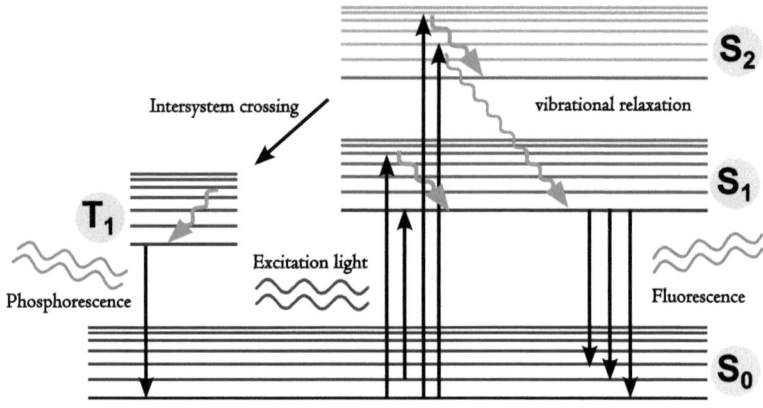

Figure 3.1: Energy level diagram after Jablonski.
The transition of an electron from the ground state (S_0) to an energetically higher level ($S_{1,2}$) is induced by the absorption of a photon. The electron relaxes back to the (S_1) state via vibrational relaxation and finally to the ground state by photon emission. This photon is visible as red-shifted fluorescence. Excited electrons can enter an excited triplet state (T_1) by intersystem crossing. Here, the energy loss is visible as long-lived phosphorescence.

3.1.2 Monochromatic light

Every fluorophore shows a distinct absorption and emission spectrum, resulting in variations of the excitation energy. Therefore, the use of a light source with an appropriate excitation wavelength can selectively excite a specific fluorophore. This enhances the signal-to-noise ratio and enables multi-color-observations by excitation with different specific wavelengths. These wavelengths can be generated from normal light by using a suitable filter system, which transmits a distinct interval of wavelengths. In contrast, a laser can produce monochromatic light and is therefore a convenient tool for most fluorescence microscopy techniques. Furthermore, lasers emit the monochromatic light as a coherent beam, which allows a spatially limited excitation with high efficiency.

3.2 Diffraction barrier of light microscopy

In order to understand, why the diffraction of light limits the resolution of a microscope, the wave model of light has to be taken into account. An obstacle in the path of a light wave causes a deflection of the light wave from its original path. This deflection is called diffraction. The diffraction of light limits the resolution of a microscope, as was first described by Ernst Abbe. This diffraction barrier depends on the wavelength and the numeric aperture of the lens and was determined as approximately half the emission wavelength [1].

The resolution limit for two point sources was further described by Lord Rayleigh. According to his work the resolution limit of two distinguishable point sources is given by:

$$d_{min} = \frac{0.61 \cdot \lambda}{\text{NA}}$$

where d_{min} describes the minimal distance between two point sources, λ the emission wavelength, NA the numeric aperture, while the factor 0.61 is derived from the distance of the first to the central maximum of the diffraction pattern. This diffraction pattern is known as the Airy disk pattern and is caused by the aperture of the lens, which diffracts the light. However, the Airy disk reflects only the intensity distribution in the focal plane, while the complete three-dimensional intensity distribution is known as the point spread function (PSF).

3.2.1 Overcoming the diffraction limit in theory

The diffraction barrier seems to prohibit the visualization of multiple nano-sized single proteins within the diffraction limit. Even the ideal aperture does not allow a better lateral resolution than 200 nm, simply because the spectrum of the visible light is limited and highly energetic wavelengths damage living systems. A theoretical consideration makes it possible to evade this problem by applying fitting methods to the PSF of the projected object. Assuming a point source of light, the maximum of an Airy disk pattern can be mathematically approximated as a Gaussian function. This makes it possible to determine the position of the point source and therefore localize the object with an up to 100 times better accuracy. Therefore, spatially well separated objects can be resolved with enhanced resolution by the application of a Gaussian fit to their PSF. The accuracy is then only dependent on the signal-to-noise ratio and is thereby limited by the number of photons emitted by the point source.

In a biological sample, the number of photons emitted by a fluorophore-tagged protein enables a resolution of about 20 nm, however the high density of interacting proteins lead to overlapping PSFs, which prevents the curve fitting as illustrated in Figure 3.2.

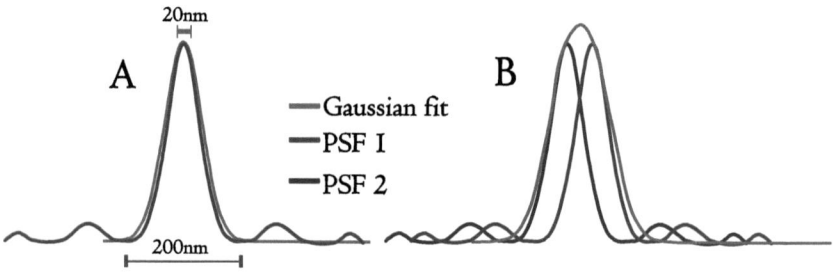

Figure 3.2: Diffraction barrier.
A single PSF can be fitted with a Gaussian curve, which allows a significantly higher localization accuracy (A). Several PSFs in close proximity cannot be distinguished from each other, which is reflected by a single Gaussian fit (B).

3.3 Super-resolution microscopy

The theory allows to circumvent the diffraction limit, but conventional fluorescence microscopy does not support a practical application. The discovery of a new class of fluorophores enabled the implementation also in reality.

3.3.1 Photoactivated localization microscopy

One prominent super-resolution technique, using photo-activatable fluorophores and their stochastic switching behavior, was developed in 2006 by three groups independently. They named it *Photoactivated Localization Microscopy* (PALM) [18], *Fluorescence Photoactivation Localization Microscopy* (FPALM) [67] and *Stochastic Optical Reconstruction Microscopy* (STORM) [119], respectively.

All three techniques share the same fundamental principle, but can be distinguished by the choice of fluorophores used and therefore in the experimental procedure. The STORM experiments were performed by using photo-switchable synthetic dyes in order to immuno-stain the endogenous protein level of the cells after fixation. PALM and FPALM were performed with photo-activatable fluorophores, which implies cell transfection and over-expression of tagged proteins. Accordingly, nowadays the term "PALM" is established for using fluorescent proteins, whereas "STORM" corresponds to the usage of synthetic dyes. The basic features of PALM to overcome the resolution limit will be explained in detail.

Photo-activatable fluorophores

Interacting proteins will always be in too close proximity to resolve with conventional fluorescence microscopy, as the average protein size is about 3-5 nm. The theoretical resolution limit breakage was not feasible for dense biological samples, until the discovery of photo-activatable (PA) fluorophores [109]. Compared to conventional fluorophores, PA fluorophores have the ability to change from a native emission state to an activated, red-shifted state upon illumination with high-energetic light. Because of their unique feature, many switchable fluorophores have been discovered in nature [155, 103] or were specially engineered [114, 64] in recent years. Nowadays, it can be distinguished between photo-activatable, -convertable and -switchable fluorophores. Photo-activatable fluorophores possess a native dark state and become visible only via activation, whereas photo-convertible fluorophores exist in a visible native emission state and change color upon activation. Photo-switchable fluorophores have the ability to switch reversibly between two states.

Generally, a conversion of color is achieved by modification of the conjugated π-electron-system within the chromophore. A conjugated π-electron-system is achieved by a high number of molecular double bonds inside a molecule, which allows the electrons to delocalize in the connected p-orbital system. An enlargement of the conjugated π-electron system shifts the color

28 *CHAPTER 3. THEORETICAL ASPECTS OF MICROSCOPY*

of the molecule to an energetically lower wavelength and can be induced by chemical reactions or conformational changes [136, 110]. The chromophore modification within irreversible PA fluorophores is usually accompanied by chemical reactions, such as decarboxylation (PA-GFP [109]) or chain breakage via dehydration reactions (mEos2 [97]), while reversible PA fluorophores function with cis/trans isomerization (Dronpa [109]).

3.3.2 Total internal reflection fluorescence

Photo-activatable fluorophores, tagged to a protein of interest enable super-resolution imaging of biological samples. However, the existence of fluorescent proteins along the z-axis reduces the signal-to-noise ratio which in turn diminishes the resolution. Therefore, another tool can be installed which allows to minimize the excitation of fluorophores along the z-axis. Total internal reflection fluorescence (TIRF) microscopy is a suitable technique, as it produces an evanescent wave of approximately 100 nanometers in the axial direction [95].

Evanescent wave

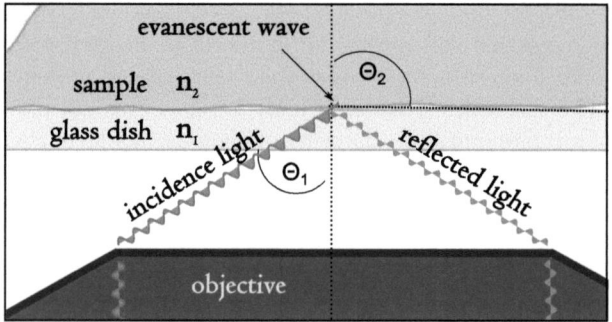

Figure 3.3: Evanescent wave.
An evanescent wave can emerge in the interface between two media with different refraction indices (n_1 and n_2). If n_2 has a lower optical density and the incidence light hits the glass slide at the critical angle Θ_1 of total reflection, an evanescent wave forms.

An evanescent wave is an electromagnetic wave that vanishes with an exponential decay and penetrates only the sample region close to the objective. There are two conditions in order to produce an evanescent field: the incidence of light must occur at an angle of total reflection and the media of the sample must have a lower index of refraction. This correlation for total reflection is postulated in Snell's law, by:

$$n_1 \sin(\Theta_1) = n_2 \sin(\Theta_2)$$

with n_1 as the refractive index of the medium where the reflection takes place and n_2 of the sample medium. Θ_1 describes the incidence angle, while Θ_2 describes the refraction angle.

With Snell's law, the angle of critical incidence can be calculated, in which the refraction angle is 90° - the angle of total reflection. A further increase beyond the critical angle will not disturb the total reflection, but the amplitude of the resulting evanescent wave will decrease. Due to the exponential decay of the evanescent wave the penetration depth is generally independent of the laser power.

3.3.3 PALM procedure

By using TIRF illumination, the fluorescence excitation along the axial direction will be limited, preventing background noise. The lateral resolution can be remarkably enhanced by tagging a protein of interest with photo-activatable fluorophores.

Here, the excitation wavelength corresponds to the color of the activated state of the PA-fluorophore. This makes all fluorophores invisible, as long as they are not activated. By using low-power activation pulses of high-energetic light (mostly near-UV light), only small fractions of the PA-fluorophores switch to the red-shifted state. During this stochastic process, the activated fluorophores tend to be spatially well separated and appear as single molecules which can be localized by applying a Gaussian fit as described in section 3.2.1.

A single activation pulse with high-energetic light enables the detection of very few molecules until they bleach, as it activates only a small subset of all present proteins. In order to obtain a complete picture, many activation cycles must be performed until all PA-fluorophores are converted. As this procedure can take some time, PALM measurements can be done only with relatively static samples.

Figure 3.4 illustrates all relevant steps of a typical PALM measurement.

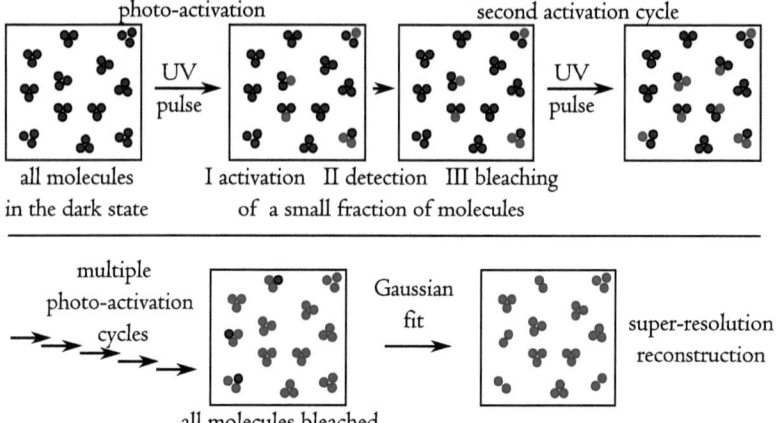

Figure 3.4: PALM procedure.
First, all molecules are invisible in the detection channel. Upon a low-power UV-pulse, very few molecules become activated and can be detected, until they bleach. This small fraction is not enough to reconstruct a complete super-resolution image, therefore, this cycle of activation and detection must be repeated until all molecules are bleached. As all molecules appear spatially separated from each other, a Gaussian fit can be applied, which enhances the resolution. The fitted localizations can be used to reconstruct a super-resolution image of the sample.

Part II
Methods

Chapter 4

General biological methods

Microscopy experiments require the preparation of suitable samples. This requires the construction of modified DNA-vectors, followed by the insertion into mammalian cells. Further, sample dishes must be purified in order to obtain a good foundation for super-resolution experiments.

4.1 DNA handling

4.1.1 Cloning strategy

All constructs mentioned in this thesis hold a cytomegalovirus (CMV) promoter for high copy amplification and are listed in Table 4.1 including their cloning details. DNA-inserts were obtained by performing polymerase chain reaction with suitable primers, using *AccuPrime Polymerase* (Invitrogen) according to the manufactor's manual. The amplified insert and the cloning vector (cf. Table 4.1) were digested by 10 U restriction enzymes (NEB), each for 2-4 h at 37°C under suitable buffer conditions. The cloning vector was additional dephosphorylated, adding 10 U CIP (NEB) for 1 h at 37°C. The digested DNA was separated by gel electrophoresis using 1% agarose-gels. The DNA-fragments were purified, using kits from *Qiagen* and *Zymo*. A ratio of 1_{vector}:5_{insert} in the femto-molar range was ligated with the *Quick Ligation* kit (NEB) for 10 min at room temperature.

4.1.2 Bacterial DNA expression

After ligation, the newly obtained DNA construct was amplified in chemically competent *Escherichia coli XL10 Gold* cells. In order to select for positive clones, the bacteria were spread on Ampicillin or Kanamycin containing media plates, according to the resistance of the cloning vector. The remaining

construct	vector	insert	enzymes	primers	vector origin
mEos2-actin**	eYFP-β-actin-ClontechC1	mEos2, mEos2-H-Ras*	AgeI, XhoI		Clontech, * Dr. Peter Verveer
β3-integrin-mEos2***	β3-Integrin-eGFP-pcDNA3.1	mEos2**	AgeI, NotI		Dr. Caroline Cruzel;
β3-integrin-mEos2; primed mutant	β3-Integrin-mEos2			CGC CTT CGA CTA CGG CCA TAT GAT TCG AAG ATC TTC TCA $C_{D1197fw}$ & GTG AGA AGA TCT GCC GCT TTA CGT^f CTG GCA TTG AAG GGC$D1197_{rv}$; CAG AGC AAA CAA CCC GGT GGC GAA AGA GGC CAC CTC CAC $C_{Y747A_{fw}}$ & GGT GGA GGT GGC CTC TTT CGC CAG CGG GTT GTT TGC TGT $G_{Y747A_{rv}}$	Dr. Peter Verveer
mEos2-Talin	Citrine-Talin-C1	mEos2	AgeI, EcoRI	mEos2-β-actin	Dr. Eli Zamir
mEos2-Kindlin2	dsRed-Kindlin2	mEos2	AgeI, XhoI	mEos2-β-actin	Prof. David Calderwood
mEos2-Kindlin2	dsRed-Kindlin2	mEos2	AgeI, XhoI	mEos2-β-actin	Prof. David Calderwood
mEos2-FAK	mtagBFP-FAK-C1	mEos2***	NheI, BglI2	CGC GCT AGC ATG AGT GCG ATT AAG CCA GAf_w & GCG CAA GAC GAT^f CCG GAC GC$_{rv}$ AAG TCC GGAλ TCG TCT GGC ATT GTC$_{rv}$	Dr. Eli Zamir
mEos2-Src	Citrine-Src-C1	mEos2***	AgeI, MfeI	CGC ACC GGT CGC CAC CAT GAG TGC GAT TAA GCf_w & GCG CAA TTG TCA TCG TCT^f GGC ATT GTC AGG C$_{rv}$	Dr. Eli Zamir
mEos2-Paxillin	Citrine-Paxillin-C1	mEos2***	NheI, BspEI	CGC GCT AGC ATG AGT GCG ATT AAG CCf_w & CTG ACA ATG CCA GAf CCG GAC GC$_{rv}$	Dr. Eli Zamir
mEos2-Vinculin	mTFP-Vinculin-C1	mEos2***	NheI, BspEI	CGC GCT AGC ATG AGT GCG ATT AAG CCf_w & CTG ACA ATG CCA GAC GAT^f CCG GAC GC$_{rv}$	Dr. Eli Zamir
mEos2-α-actinin	mtagBFP-α-actinin-C1	mEos2***	AgeI, NotI	CGC ACC GGT CGC CAC CAT GAG TGC GAT TAA GCf_w & GCG GCC GCT TTA CGT^f CTG GCA TTG	Dr. Eli Zamir
mEos2-Zyxin	Citrine-Zyxin-C1	mEos2***	NheI, EcoRI	CGC GCT AGC ATG AGT GCG ATT AAG CCA GAf_w & GCG GAA TTC TCT CGA GAT CTC CCT CGT CTG GCA CTC	Dr. Eli Zamir
mEos2-TMD	PAmCherry-TMD-C1	mEos2***	AgeI, NotI	CGC ACC GGT CGC CAC CAT GAG TGC GAT TAA GCf_w & GCG GCC GCT TTA CGT^f CTG GCA TTG	Dr. Peter Verveer
mtagBFP-Zyxin					Dr. Eli Zamir

Table 4.1: **Cloned constructs.** The asterisks mark the origin of the insert or construct.

4.2. CELL CULTURE

bacterial colonies were screened for potential positive clones, using *Paq5000* (Agilent Technologies) and after that sequenced with the *BigDye* sequencing kit (Life Technologies).

4.1.3 Site-directed mutagenesis

For the induction of point-mutations in DNA constructs, modified primers were used containing the desired mutation. The template DNA was amplified with *Pfu Ultra* (Agilent Technologies) by performing polymerase chain reaction, using these primers. Additionally, 5% DMSO was added to the reaction mix. The modified DNA construct was amplified in *Escherichia coli XL10 Gold*.

4.2 Cell culture

For PALM experiments, rat embryo fibroblast (REF52) cells were used because they produce a high amount of extracellular matrix proteins, thus forming well defined focal contact sites. For live cell experiments, the epithelial human cancer cell line HeLa was used, which shows generally more robustness towards long-term measurements.

4.2.1 General cell line handling

Both cell lines were grown in 10 ml-culture flasks containing DMEM, supplemented with 10% fetal calf serum, 1% glutamine, 1% non-essential-amino acids and 1% antibiotics (DMEM$_{complete}$) and incubated at 37°C under 5% CO_2 conditions. When a cell confluency of 80-90% was reached, cell splitting was performed. For that purpose, DMEM$_{complete}$ was removed, the cells were washed once with PBS and incubated with trypsin/EDTA until all cells detached from the dish. To stop trypsin's enzymatic activity, DMEM$_{complete}$ was added and 10% of this cell suspension re-seeded into a new 10 ml-culture flask, filled with DMEM$_{complete}$. For experimental use, 2.5×10^4 REF52 or 5×10^4 HeLa cells were seeded into each well of a PAA 8-well-chambers and grown in DMEM$_{complete}$. If the cells were not needed for longer time, they could be stored in liquid nitrogen. Therefore, $1\text{-}5 \times 10^6$ cells were placed into a vial with 500 µl DMEM$_{complete}$, containing 10% dimethyl sulfoxide (DMSO) and stored at -80°C for 2 days until passed on to a -196°C freezer for long-term-storage. For further use, the frozen cells were placed into a 10 ml DMEM$_{complete}$ containing culture flask and incubated at 37°C and 5% CO_2. After initial cell adhesion (after 1-2 h), a media change was performed, as DMSO leads to cell toxicity under cell culture conditions.

4.2.2 Transient transfection

PALM experiments required the transfection of a single protein, tagged with the photo-convertable mEos2. In live cell experiment, the mechano-sensor Zyxin was co-transfected in order to distinguish between focal complexes and focal adhesions. 12-18 h after seeding, the cells were transiently transfected with *Lipofectamine 2000* (LifeTechnologies). For single transfections, 300 ng DNA and 1 µl Lipofectamine per well were used. For double-transfections with mtagBFP-Zyxin and a mEos2-construct, a ratio of 2:1 was used while the amount of Lipofectamine remained the same. The DNA-Lipofectamine-complex could form in DMEM for 30 min at room temperature, before adding it to the cells. The transfected cells were incubated for 24-32 h together with the transfection reagent.

4.3 Experimental cell preparation

4.3.1 Preparation of cell dishes for microscopy

For all experiments 8-well *LabTek* chambers (Thermo Fisher Scientific) were used. Due to cleaning issues, the chambers were incubated with 1% *Hellmanex III* (Hellma) for 12h and subsequently washed 2x with ddH$_2$O.

For live cell recruitment experiments, the cleaned *LabTek* chambers were coated with the ECM protein Vitronectin (Sigma Aldrich). Vitronectin is recognized by β3-integrin and provides a stable anchorage of the cell. Cell migration is rather suppressed, which enhances the quality of live cell experiments in terms of localization precision. Therefore, a solution of 0.5 g/ml Vitronectin in HBSS was plated and incubated for 2h at 37°C and 5% CO$_2$. Afterwards, the dishes were washed 3x with HBSS.

To avoid high background, caused by sediments of media additives and cellular outturn, the transfected cells were re-seeded into a prepared *LabTek*-chamber only 12-16 h before the planned experiment. For live cell integrin cluster induction experiments, HeLa cells were incubated in serum- and ion-free PBS before the experiment. All other experiments required no special treatment and cells were grown in DMEM$_{complete}$. 12-16h after re-seeding, HeLa-cells were used for microscopy, whereas REF52-cells were further proceeded.

4.3.2 Fixation and staining of REF52 cells

Fixed cells are mandatory for PALM experiments, as protein dynamics in live cells would hamper PALM experiments. Transfected REF52 cells were washed 3x with PBS before fixation in 4% PFA for 1 h at room temperature. After washing another 3x with PBS the cells were either used directly for microscopy or additionally stained with a $\text{FAK}^{pTyr397}$ antibody (44-624G, Invitrogen). $\text{FAK}^{pTyr397}$ staining is required as a marker for control samples, that over-express proteins which do not localize in focal adhesions. A successful staining includes an incubation with 0.2% TritonX100 for 10 min in order to permealize the membrane. After washing the dish 3x with PBS for 5 min each time, a 1:100 solution of the primary antibody FAK^{397} in PBS + 0.02% TritonX100 (PBS_{TX}) was incubated for 1 h at room temperature. After washing 3x for 5 min with PBS_{TX} a 1:1000 solution of secondary goat Alexa647-anti-rabbit antibody (Invitrogen) in PBS_{TX} was incubated for 1 h at room temperature. The cells were washed another 3x with PBS_{TX} for 5 min at a time, followed by microscopy experiments.

4.3.3 Inhibition of mechanical force

The formation of F-actin stress fibers is inhibited by the ATP-competitor Y-27632 (Sigma Aldrich) [143], which is selective for the Rho downstream effector ROCK (Rho-associated coiled coil forming protein serine/threonine kinase). Incubation with 10 µl Y-27632 causes a reversible loss of adhesion sites and was implied for 30 min in PBS before fixing the cells. For live cell experiments, the drug was added during measurement and incubated until the adhesion marker protein Zyxin left the focal complex completely. This took usually up to 30 minutes. Afterwards, the drug was removed in three HBSS washing cycles.

CytochalasinD binds to the growing end of F-actin and thereby stops any further actin polymerization. As it does not affect the de-polymerization of the shrinking actin end, it supports the disassembly of stress fibers [50, 147]. This impedes the formation of focal adhesions and causes also their dissassembly. REF52 cells were incubated for 1 h with 5 µM CytochalasinD by HBSS before fixation. No live cell experiments were performed with this drug.

4.3.4 Induction of Integrin clustering

The formation of new adhesion sites was observed in living HeLa cells. The activation state of integrin receptors is highly dependent on the cation environment, as explained in chapter 1.2.1. Non-physiological Manganese ions induce the high affinity state of integrin receptors [137], followed by cluster formation. In order to artificially induce this first step of adhesion sites formation, $2\,\mu M$ $MnCl_2$ in HBSS containing $1\,mM$ Mg^{2+} and $1\,mM$ Ca^{2+}, for synergistic support was added to starved cells. The added media contained also 5% fetal calf serum, which triggers migration and allows the formation of mature focal adhesion sites.

Chapter 5
Data acquisition & analysis

All experiments were performed with an Olympus IX 81 (TIRF) microscope equipped with a 60x TIRF oil objective. For PALM experiments, an objective with NA 1.49 (Olympus) was used; for live cell experiments an objective with NA 1.45 (Olympus) equipped with a heating device (Chromaphore). The images were recorded with an Andor DU897bV camera, which amplifies the detected signal by an electron multiplying charge-coupled device (EM-CCD).

As light sources served an Argon laser (Coherent) emitting at 476 nm, a Krypton laser (Coherent), emitting at 568 nm, a diode laser emitting at 405 nm (UV) as the photo-activation provider and a diode laser emitting at 641 nm (far red) as antibody staining control.

The set-up is schematically illustrated in Figure 5.1. Both diode lasers are equipped with optical density filters in order to prevent high laser power exposure. The Krypton and the far-red laser are coupled to the same glass fiber, but can be used individually. The microscopic aperture of the UV laser regulates the area of activation and can be closed, if only a small area should be activated. In the microscope, all laser beams are directed to the TIRF objective, where they hit the sample. Emission and excitation light are separated by a dichroic mirror before the emission light is detected by the EM-CCD.

5.1 Photoactivated localization microscopy

For PALM experiments, the sample has to be fixed (cf. chapter 4.3.2), as all single molecule detections are acquired over a long time, as explained in chapter 3.3.3. For super-resolution imaging, the photo-convertible fluorophore mEos2 was used. Due to its green fluorescence before photo-conversion, a control in the green channel was done before and after the measurement.

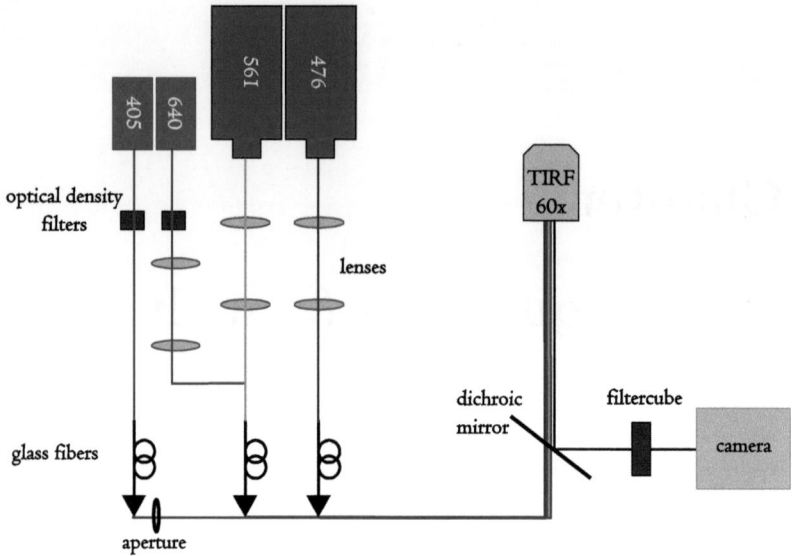

Figure 5.1: Super-resolution imaging set-up.
All experiments were performed with this schematically illustrated set-up. The Krypton and the far-red laser share the same glass fiber. In the microscope, all laser beams are combined and directed to the TIRF objective and focused in its back focal plane. Furthermore, the TIRF objective collects the emitted fluorescence light. A dichroic mirror separates the emission light from the excitation light and directs it to the EM-CCD, passing a suitable filter.
During PALM experiments, an optical density filter of 0.1 for the UV laser and 0.05 for the far-red-laser were used. For live cell spot activation, the aperture of the UV-laser was closed. For all other experiments, it was open.

Hence, a pre-check of transfection efficiency and protein localization and a post-check of successful photo-conversion is provided. mEos2 activation was done by UV light exposures, whereby the time points and time spans were chosen manually. FAK-stained samples could be imaged in the far red channel.

5.1.1 Protocol of data acquisition

Prior to the imaging of the samples, the lasers and the TIRF field were aligned. First, the Argon laser was used at low powers of 2-10 mW to search

5.1. PHOTOACTIVATED LOCALIZATION MICROSCOPY

for a suitable cell. When Alexa647-immuno-stained cells were used, the far red channel was imaged first. A power of 0.6-2.4 mW and an additional optical density filter of 0.1 was sufficient and prevented overexposure and bleaching. Afterwards, the sample was illuminated with the Krypton-laser until all unspecific particles were bleached to obtain a super-resolution image with minimal artifacts and background. Then, the recording of single molecules started, using an exposure time of 100 ms (10 frames/sec).

When performing PALM experiments, ideally all molecules are detected only once and are visible for only one frame to avoid multiple detections. On this account, the mEos2 read-out laser was used at its maximum power of 100 mW. The UV laser was manually switched on and off, according to the amount of visible molecules. The applied UV laser power was highly depending on the expression level and localization of the mEos2-tagged protein and started at 0.1 mW, whereby an optical density filter (0.1) was used (resulting in a power of 0.01 mW) and went up to 10 mW. Recording was performed until no specific detections were observed, which took usually 4000-20000 frames and was confirmed by taking an image of the Argon channel. If no fluorescence was visible there, all molecules were converted.

5.1.2 Data analysis

In order to obtain information about the nano-organization of molecules within a cell, several processing steps are required. Starting from a general transformation of image to xy-localization information (RapidStorm), continuing to the selection of regions of interest (in MATLAB) followed by the analysis of the organization degree (based on degree distribution). Finally, the dense structures were analyzed, aided by *density-based spatial clustering of applications with noise* (DBSCAN) [47].

RapidStorm

In order to convert the imaged frames of the measurement to a data table with all localizations, the software *rapidStorm* (version 2.21) [158] was used. Each dataset contained up to 4086 frames and was analyzed with an estimated PSF-FWHM (full width at half maximum of the optical point spread function) of 250 nm, based on the emission wavelength. The output pixel size size of 107 nm was determined by imaging a standardized micrometer target. Furthermore, a threshold of 1000 photon counts per localization was applied to all datasets, as well as an *Two-Kernel*-improvement. A *Two-Kernel*-analysis applies a Gaussian fit to the PSFs, consisting of two Gaussian functions. Therefore, the amplitudes, derived from the single molecule detections will be screened for potential double detections. *Rapidstorm 2.21* is already capable of merging multiple consecutive detections of the same molecule into a single position, by identifying a single fluorescent molecule over consecutive frames.

Artifacts of the measurement

Even though the UV-activation in a running experiment was balanced well, in a dense environment like adhesion sites several molecules in direct proximity can be activated at the same time. Two-Kernel-analysis can correct such cases only if the PSF of two molecules overlap completely. In other cases, such detections will be considered as noise.

A long-lived or blinking fluorophore can induce multiple detections, resulting in a false cluster. Even though, this is taken into account by the analysis, photon emission fluctuation or an extended triplet state can induce false localizations. Furthermore, some molecules are not considered, because they emit too few photons (below 1000 photons), or do not photo-convert at all.

Region of interest selection

The localizations of one image, determined with *Rapidstorm 2.21* can be divided into several output-files, due to memory storage reasons. Such localization files can be merged with a custom-made MATLAB script. Furthermore, a second data cleaning step was applied, using the same parameters, as in RapidStorm, but taking the triplet state into account. In rare cases of intersystem crossing (cf. chapter 3.1.1) the emission can take longer than 100 ms (1 frame), which is considered in this MATLAB computation with 3 frames (300 ms). Then a super-resolution image is generated by plotting the localizations.

The number of localizations of a whole cell is too big for further calculations. Therefore, only regions of interest (ROIs) were selected. This was done manually, by choosing rectangular ROIs of focal adhesion sites of the cell body and the leading edge, as well as membrane areas without visible focal contact sites. Only these selected localizations were further analyzed.

Density determination

As an indicator for the expression level, the density of each focal adhesion, selected in section 5.1.2 was determined. Therefore, a broadened Gaussian blur was applied to all localizations. As a result each localization covered a circular area with a diameter of about 100 nm, which was used as a mask for

5.1. PHOTOACTIVATED LOCALIZATION MICROSCOPY

the total area. The total number of localizations was divided by the total area and represents the average density of each ROI.

Degree distribution

The localizations of the selected ROIs were used to get a deeper insight into the nanometric structure of each observed adhesion protein. Therefore, the number of connections of each molecule to neighboring molecules was counted using degree distribution analysis [40]. This algorithm is generally used for network reconstruction and counts the number of connections for each network node, which is defined as the degree. To obtain a probability distribution of the connections, the degree is divided by the number of molecules and can be plotted against the connections. This histogram represents the degree distribution and mirrors the average connections in each network, peaking at the most probable one.

Degree distribution can be implemented on localizations, obtained by super-resolution imaging and unravel information about the nano-scale organization. Therefore, each localization was treated as a node. The network was artificially defined by a radius, in which the number of localizations were counted. To do so, an adjacency matrix A was calculated, which contains the matrix elements a_{ij}, representing the connection between molecule i and molecule j. If two molecules are closer together than the chosen radius, the particular matrix element is set to 1. Otherwise, the value is set to 0.

$$A = \begin{pmatrix} a_{1,1} & . & . & a_{1,j} \\ . & . & . & . \\ . & . & . & . \\ a_{i,1} & . & . & a_{i,j} \end{pmatrix}$$

The degree d_i of the molecule i is defined as the sum of elements along row i. The degree distribution is then the fraction of nodes (molecules) with their calculated degree d:

$$d_i = \sum_j a_{ij}$$

The difficulty here is to chose a suitable radius. Some proteins might interact directly, which requires a rather low radius. Other proteins might organize on a higher level and require a higher radius. Considering this, a range of different radii was calculated, starting from 20 nm limited by the resolution of the microscope and moved up to 50 nm, which corresponds to the maximal ligand-spacing distance for successful focal adhesion formation [10, 28].

Application of degree distribution on the lateral positions revealed localizations in close proximity, which corresponds to protein aggregations. A radius of 25 nm was applied in order to detect such clustered regions in a lateral reconstruction of the localizations in fixed cells as will be further described in chapter . In living cells, a radius of 150 nm was applied which will be explained in chapter .

Cluster definition

Such dense areas were further analyzed, using a MATLAB implementation of DBSCAN (density-based spatial clustering of applications with noise) [41]. As indicated by the name, DBSCAN defines proximal localizations as a cluster by density reachability. Also here, a radius must be given in order to define a density reachability, as well as a number for the minimal localizations forming a cluster. Due to the large variations in size and density of protein aggregates, DBSCAN could not deliver suitable results. As an alternative, DBSCAN was only used for detecting and characterizing pre-defined clusters. A cluster was defined, according to its number of connections, obtained by degree distribution.

Due to the large variations in the number of connections of each cluster, it was mandatory to manually choose an individual threshold of the number of connections for each focal adhesion site. Localizations, with more connections than 65% of the applied threshold belonged to very dense regions. Therefore, only localizations of this upper third were used for further analysis. The reduction of localizations enabled DBSCAN to trace clusters, which were further fitted using a custom-written MATLAB script. Not all clusters were round-shaped, so an elliptical fit was applied. The area of the clusters as well as the axial radii of the elliptical fit were calculated. Furthermore, the standard deviation s was calculated by:

$$s = \sqrt{\frac{1}{n-1}\sum_{i=1}^{n}(x_i - \bar{x})^2}$$

whereby n represents the total number of datasets, x_i is an individual dataset and $\bar{x}$ corresponds to the mean value of all datasets.

5.2 Single particle tracking

When using living cells, a TIRF objective was used, equipped with a heating device, set to 37.5°C. For live cell experiments, an exposure time of 31.6 ms

5.2. SINGLE PARTICLE TRACKING

was used, resulting in 30 frames per second due to the read-out time of the camera. Contrary to PALM, it is important to reduce bleaching of the sample to gain the longest possible trajectories during *single particle tracking* (SPT). Hence, the Krypton laser power was set to 10-20 mW.

5.2.1 Data aquisition for β3-integrin tracking

For SPT-experiments, usually mEos2-transfected HeLa cells, co-transfected with mtagBFP-Zyxin were used. Then, a suitable cell was searched in the Argon channel. Before the actual measurement, the Krypton channel was bleached in order to minimize unspecific detections.

At first, an image of the blue channel was taken with the UV laser at about 5 mW, directly followed by SPT in the Krypton channel, as the UV laser induced the photo-conversion of mEos2-tagged proteins. During the acquisition of one dataset of 2000-4000 frames (1-2 min) no further exposure to UV light took place. This repetitive cycle was performed throughout the experiment.

5.2.2 Data acquisition of spot activation

For recruitment experiments, only mEos2-tagged proteins in a defined part of the cell were photo-converted, which could be realized by closing the aperture within the beam path of the UV laser. The position of the UV spot is defined by the aperture and cannot be changed. The area exposed to the UV-light cannot be considered in the data analysis, which excludes all adhesion sites in this region. Therefore, the cellular position must exhibit a good amount of well localized adhesion sites outside the activation area. The adequate positioning can be done in the Argon channel.

Before the measurement, an obligatory bleaching step took place until only random blinking was visible, which occurred due to spontaneous photo-conversion of a sub-fraction of mEos2 upon illumination with low-energetic light.

A control dataset of 2000 frames was taken with random blinking before two consecutive datasets of 2000 frames each, were recorded. Right in the beginning of one dataset the UV laser was switched on for 10 sec at 5 mW. This time, the UV laser was in wide field mode instead of TIRF to ensure the illumination of the cytosolic fraction.

5.2.3 Data analysis

In order to detect trajectories, single molecule localizations must be extracted from the raw data, fitted with a Gaussian and further connected to whole trajectories depending on their proximity in consecutive frames. This was done by *u-Track* - a software, specialized for data with high particle density [73] and therefore suitable for investigations in dense domains like focal adhesion sites.

Afterwards, the diffusion constant D can be extracted for each track, using a mean square displacement (MSD) analysis on the base of a Brownian motion model [100]. MSD calculates the distance, that a particle moves within a given time Δt (here: one frame to the next in 31 ms).

$$\text{MSD} = (\epsilon - \frac{4}{3}Dt_e) + 4D\Delta t$$

with $\epsilon = 4\,\sigma$ and σ as the Gaussian-distributed localization uncertainty. σ is derived by determining 15 MSD points of each trajectory. Then, a linear fit was applied to this curve using 4 points. Each trajectory must exceed a defined length of 5 frames (0.2 seconds) in order to be considered for MSD analysis.

In some cases, only determined regions, like focal adhesions, were observed. Therefore, a suitable mask was produced in *ImageJ* [3] and applied during MSD calculation.

Furthermore, general normalization and standard deviation calculations according to section 5.1.2 were applied to the diffusion constants.

Part III

Results

Chapter 6

Super-resolution imaging of adhesion sites

The static position of a focal adhesion is achieved by the relatively immobile binding of integrin receptors to their extracellular ligands. All cytosolic proteins assemble to this initial anchoring. Hence, a spatially well structured organization of focal adhesion sites was hypothesized [12, 55, 131]. This hypothesis is further supported by super-resolution images [76, 117, 33], showing a distinct organization of adhesion sites, but mainly in axial than in lateral direction.

In Figure 6.1 a typical PALM image of the mEos2-tagged scaffolding protein Zyxin is shown. The molecules seem to form equally distributed and laterally defined clusters. This molecular composition can be analyzed and compared with other proteins, involved in a functional focal adhesion in order to unravel the individual lateral nano-organization of each protein.

The receptor β3-integrin, the first binding partners Talin, Kindlin1, Kindlin2, the signaling complex components FAK, Src, Paxillin, the scaffolding proteins Vinculin, Zyxin, α-actinin and the cytoskeleton builder β-actin were imaged by super-resolution microscopy according to section 5.1.1 and section 5.1.2 of chapter 5. A randomly distributing α-helical transmembrane domain with cytosolic mEos2-tag (TMD) was also measured, serving as a reference.

6.1 Spatial organization

The lateral localizations of adhesion proteins, obtained by PALM experiments were analyzed according to their molecular proximity using degree distribution analysis (cf. chapter 5.1.2). The degree distributions of one particular protein were merged together into a 3-dimensional plot, showing the

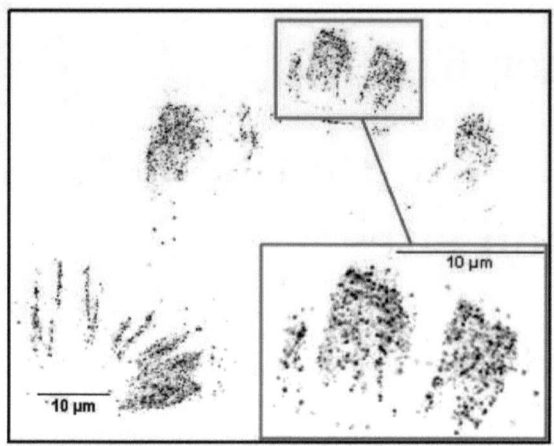

Figure 6.1: PALM image of mEos2-Zyxin.
This super-resolution reconstruction of Zyxin suggests a rather symmetric organization, in which clustered areas alternate with non-clustered areas.

number of neighbors of each detected molecule (x-axis) versus a nanometer range of 20-50 nm (y-axis). This range corresponds to studies of the adhesion ability in terms of integrin ligand spacing, which revealed a maximum distance of ligands below 60 nm for a successful cell adhesion [10, 28]. The third dimension corresponds to the probability and is encoded by color, whereby red represents the highest probability and blue the lowest.

6.1.1 Degree distribution of simulations

Three parameters can be derived from degree distribution analysis, which are visualized in Figure 6.2A on the basis of a simulated random distribution. First, the distribution with the highest probability (peak value), which states the most probable number of molecules in a cluster and the corresponding nanometer size of such a cluster. Second, the width of the most probable composition in terms of connections can be defined using the full width at half maximum (peak width). A narrow distribution would correspond to a more defined organization of the proteins within the observed radii-range of

6.1. SPATIAL ORGANIZATION

20-50 nm. Third, the inclination of each protein distribution is defined by drawing a diagonal with an angle of 45° from the peak value. This diagonal defines the upper and the lower area of the degree distribution. Then, the particular probabilities in both areas are calculated, followed by determining the ratio of upper and lower area. Simulated distributions reveal a rather high ratio for random distributions and a low ratio for highly structured distributions.

Figure 6.2 illustrates further the degree distributions of two simulations, with defined clusters. In Figure 6.2B a 4 nm cluster containing 4 particles is formed, whereas Figure 6.2C represents a 50 nm cluster, containing 100 particles. The x-axis of both plots was moved by +1 and represents the particles instead of the neighbors, for a better understanding.

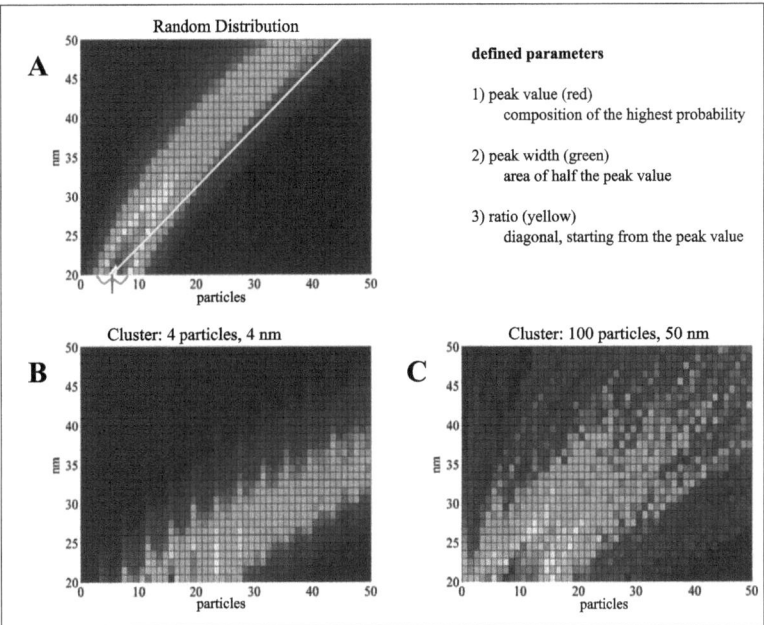

Figure 6.2: Degree distribution of simulations.
The degree distribution of a random distribution is shown in A. In this example, the derivable parameters are demonstrated and explained. The degree distribution of another simulation, shown in B has defined maxima, which allow the prediction of a 4 nm cluster, consisting of 4 particles. No prediction can be derived from degree distribution C, which represents a simulation of 50 nm clusters with 100 particles, each.

The distribution is clearly distinguishable, as the defined clusters in Figure 6.2B exhibit sharp maxima in a step size of 4 particles, each. The lowest y-value is per definition 20 nm and corresponds to the maximal peak value, containing 16 particles. However, the 3-dimensional degree distribution reveals also smaller organization with lower probability. Even well below resolution limit, a fainting peak can be observed corresponding to 4 particles.

Larger clusters of intrinsic randomness produce a different degree distribution, as observable in Figure 6.2C. Here, the distribution does not peak at distinct values. Instead, the distribution allows no conclusions about potential clusters, since the cluster size coincides with the maximal nanometer value of 50 nm. Particles within this simulated cluster are not organized, thus resembling a random distribution. Still, Figure 6.2C does show alterations in the degree distribution, compared to a complete random distribution shown in Figure 6.2A. Therefore, larger clusters with a random intrinsic organization cannot be well defined by degree distribution; though they are distinguishable from complete random organizations.

6.1.2 Degree distribution of adhesion proteins

The degree distributions, obtained from real PALM data of adhesion proteins can be interpreted on the base of the simulations in Figure 6.2. Figure 6.3 illustrates such degree distributions and Table 6.1 lists the derived parameters. In all illustrated degree distributions the highest probability (colorbar) was set to 10% for better comparison.

Generally, all degree distributions in Figure 6.3 exhibit a rather narrow distribution. Only a single peak value with a broad peak width can be observed, which contradicts a distinct clustering on a nanometer-level (cf. Figure 6.2B). However, none of the distributions corresponds to a complete random distribution (cf. Figure 6.2A). Therefore, the tendency of adhesion proteins to cluster must lie somewhere in between.

The non-functional TMD should distribute randomly in the cell membrane, whereas its degree distribution exhibits a non-random organization (cf. Figure 6.2A and Figure 6.3). This observation indicates that transmembrane proteins do have the tendency to form complexes, even though they have no biological function. The distribution of Kindlin1 is nearly identical to the randomly distributed TMD, which corresponds to the observation that Kindlin1 is distributed across the entire plasma membrane, including focal adhesion sites. Thereby, an equal distribution of Kindlin1 was observed across the whole membrane, including focal adhesion sites. In this respect, it is worth mentioning that Kindlin1 is a cytosolic protein, whereas TMD spans the membrane. Still, both degree distributions match nearly perfectly, which

6.1. SPATIAL ORGANIZATION

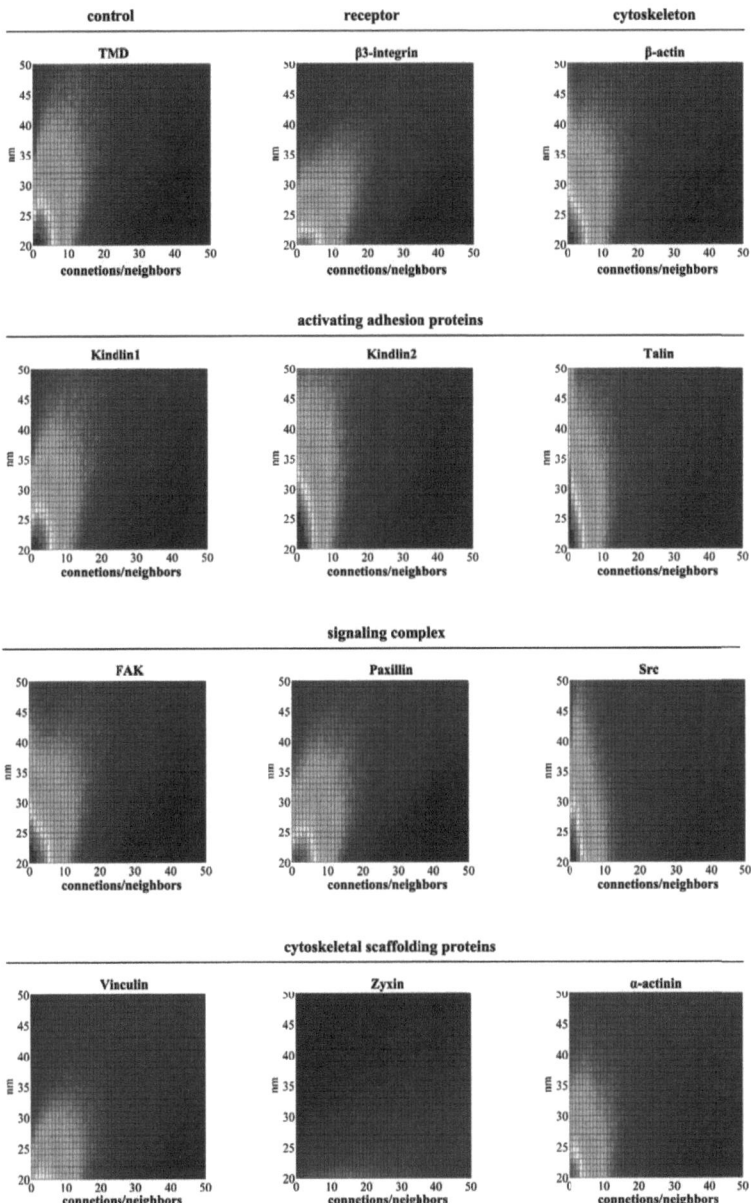

Figure 6.3: Degree distribution of adhesion proteins.
The degree distributions exhibit only one peak value and mostly a broad distribution width, which contradicts a highly organized system. A tight and defined clustering would reveal a similar pattern as in Figure 6.2B, which suggests that none of the investigated adhesion proteins form such clusters. Instead, larger clusters could be formed with an individual intrinsic distribu-

validates the use of TMD as a control for receptors as well as for cytosolic proteins.

No adhesion protein seems to organize in a distinct pattern, which is conserved over multiple adhesion sites. Still, they do not match a random distribution of membrane proximal proteins like TMD and Kindlin1, either. Some cell-matrix proteins even show similarities in their probability distribution.

Talin and Kindlin2, as well as Src seem to have surprisingly few neighboring molecules within a 20-50 nm range, with a comparatively distinct peak distribution (cf. Table 6.1). In reverse this means that on average, these proteins tend to keep larger distances and do not cluster extensively in a radius up to 50 nm. Especially Talin and Kindlin2 exhibit a distinct peaking at 1 or 2 for each chosen radius, which corresponds to maximal one neighbor on average. Similarities in the spatial organization of Talin and Kindlin2 are reasonable, as these proteins were shown to have a synergetic effect in terms of receptor activation (cf. chapter 1.2.1). This suggests a tight interaction on the molecular level.

Src exhibits a similar distribution, even though it is a major player of the signaling complex and does not interact with Talin or Kindlin2 directly. Instead, Src forms a complex with FAK and localizes tightly with Paxillin (cf. chapter 1.3.1), which is neither mirrored in Figure 6.3 nor Table 6.1. FAK shows a degree distribution, which resembles rather the degree distribution of the TMD in terms of the calculated parameters in Table 6.1. The degree distribution and the calculated parameters of Paxillin suggest a rather individual organization, without similarities to any of the other cell-matrix proteins. The cytoskeletal scaffolding protein Vinculin has no clearly favored distribution that could be distinguished with a probability of 10% (setting of the colorbar). A reason for this observation could be Vinculin's multiple binding sites in the rod-domain of Talin, which was shown to span an axial area of about 40 nm [76]. Since a fully extended Talin can span about 60 nm (cf. Chapter 1.2.1), Talin accumulates most probably in an angled position. Therefore, also Vinculin accumulates multiple times in axial and lateral direction, which complicates a proper lateral resolution. The distribution of Zyxin is even more unspecific, as it spans a total distribution of nearly 80 neighbors (cf. Table 6.1). No preferred cluster degree can be defined pointing to a broad and unresolvable spatial distribution caused by Zyxin's multiple interactions within the actin stress fibers, which are spanning the whole cell in axial and lateral direction.

α-actinin is a cytoskeletal scaffolding protein too, but reveals a different degree distribution than Vinculin and Zyxin. According to Table 6.1 the nanometric organization of α-actinin is rather individual. Furthermore, the

6.1. SPATIAL ORGANIZATION

Protein	Ratio	y (radius in nm)	x (neighbors)	Peak width
Actin	1.4415	20	1	6
α-actinin	1.2068	20	1	8
β3-integrin	1.0199	20	1	11
β3-integrin$_{primed}$	1.2128	20	1	9
FAK	1.3255	20	1	7
Kindlin1	1.3293	20	1	8
Kindlin2	1.8628	20	1	5
Paxillin	1.2802	21	2	10
Src	1.4543	20	1	5
Talin	1.7233	20	1	4
TMD	1.3548	20	1	8
Vinculin	0.9079	22	3	14
Zyxin	0.7071	35	15	38

Table 6.1: Calculated parameters of the degree distribution of focal adhesion proteins.
x and y correspond to the peak value.

degree distribution of Actin does not show significant similarities to any of the other proteins.

All cytosolic proteins could theoretically accumulate in axial direction, which would produce denser areas like the simulation in Figure 6.2C. If these axial accumulations are randomly distributed or fluctuate highly in their composition, it will be reflected as a broad peak width in the degree distribution. Compared to Figure 6.2C, the peak width is rather small for most proteins. Therefore, adhesion proteins could form large and dense accumulations, but the individual degree distributions could indicate a structured organization inside such large accumulations.

Degree distribution of β3-integrin

The only functional protein accumulating exclusively in lateral direction is represented by β3-integrin. However, Table 6.1 reveals that also β3-integrin seems to build no defined cluster pattern within a 50 nm radius, as was observed for all other adhesion proteins. Therefore, a potential structured organization of adhesion proteins cannot be masked by their axial accumulation. Still, a structured accumulation in larger clusters is possible, also for β3-integrin. Assuming that β3-integrin does have a structured organization, but in a way that is not easily resolvable with degree distribution, the degree distribution of a mutated form of β3-integrin could exhibit differences in terms of neighborhood and favored composition.

The activation of integrin receptors occurs bidirectional (cf. Chapter 1.2.1). A different activation mechanism could entail a different molecular distribution. As already mentioned, Figure 6.3 represents an average degree distribution, including many datasets of different focal adhesions. It is not possible to predict, which activation mechanism was responsible for their cluster organization. Therefore, a mutated integrin was designed, which only reacts on inside-out-activation (cf. Chapter 4.1.3) primed by Talin. β3-integrin$_{primed}$ has two mutations (D119Y and D747R), which prevent ligand binding and disrupt the cytosolic saltbrigde formation of the heterodimer. Therefore, outside-in activation is blocked and inside-out-activation supported, because Talin can access the cytosolic tail of the β-subunit easier.

The mutant β3-integrin$_{primed}$ reveals a slightly different distribution compared to the wild type β3-integrin, as is illustrated in Figure 6.4. Also Table 6.1 reflects such changes in a higher value for the ratio and a more distinct peak width, although the peak value remains the same. β3-integrin$_{primed}$ seems to have less neighbors on average, which could be interpreted as a different molecular patterning compared to the wild type.

However, it is questionable, whether such small changes can be considered as an altered nanometric organization. The mutant might also localize in focal adhesions, formed by outside-in signaling. However, due to its ligand binding inability, its total amount in such adhesion sites might be smaller. As a result, the average cluster size would shrink.

This suggests further, that the expression level of each protein could influence the average degree distribution significantly, as less clustering is observed in a low transfected cell.

6.1.3 Influence of the expression level

The results obtained by degree distribution analysis exhibit no distinct nanometric cluster formation of adhesion proteins below 50 nm. However, most proteins do show an individual probability composition, that differs from a random distribution. Whether cell-matrix proteins contain a conserved individual molecular structure, and to which degree, cannot be answered by these results.

The cellular expression level could be a key criterion for a broad degree distribution. In a poorly transfected cell, a high level of endogenous proteins interfuse the focal adhesion. mEos2-tagged proteins would cluster with the endogenous pool in order to scaffold the focal adhesion, which causes a fragmentary picture of the composition in a PALM experiment. In a focal

6.1. SPATIAL ORGANIZATION

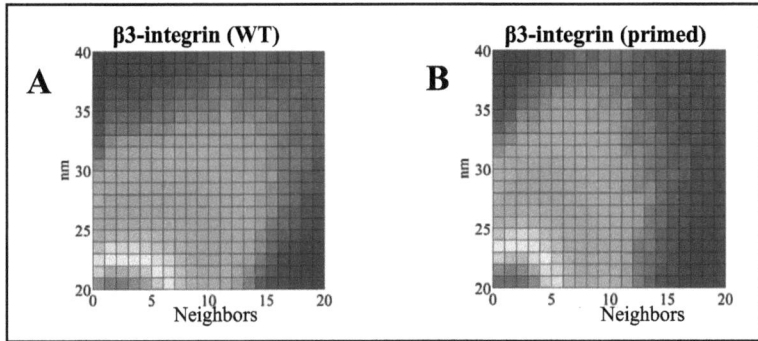

Figure 6.4: Comparison of the degree distributions of the wild type β3-integrin and a Talin-primed mutant β3-integrin$_{primed}$. Both integrins have a similar degree distribution, however the peak of the mutant (B) is minimally shifted towards less neighbors. This could indicate a different intrinsic organization of β3-integrin$_{primed}$ compared with the wild type in adhesion sites.

adhesion with high endogenous protein level, the over-expressed, activatable fraction will deliver a super-resolution reconstruction, which has less proteins in a cluster, meaning less molecules in close proximity. If these localizations are mixed with localizations derived from cells with a lower endogenous level in the adhesion sites, a degree distribution with less defined peaks would be the result. The peak value could be shifted towards less clustering and the peak width would increase dramatically.

For PALM experiments only cells with a high transfection level were chosen to ensure an almost complete occupancy of tagged proteins in focal adhesions. This could be verified by a high cytosolic pool of over-expressed protein. Nevertheless, a further control of the expression level could help to interpret the degree distributions.

The expression level corresponds to the density of the localizations inside a focal adhesion, which was calculated after Chapter 5.1.2. Significant variations in terms of expression level between cells should be visible by different density of their focal adhesions. In Figure 6.5 the densities of all focal adhesions analyzed in section 6.1 were plotted according to their overexpressed protein. Only the scaffolding proteins Vinculin and Zyxin, that possibly localize axially in multiple layers, reveal a comparably broad spectrum of different densities. All other proteins show a tighter density range, though not as narrow as the TMD control. The question remains, whether such variations are caused by different expression levels or rather by a differ-

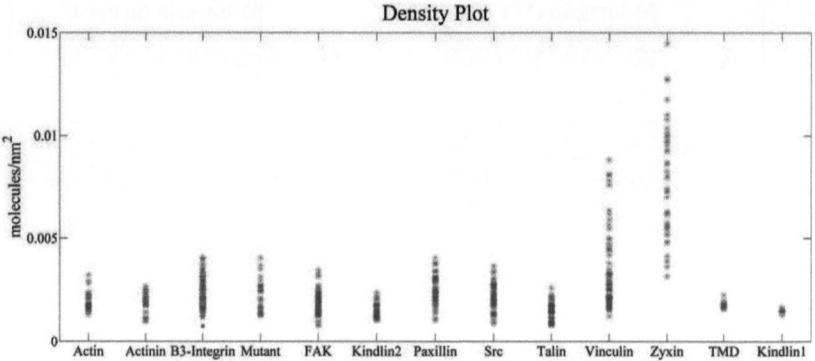

Figure 6.5: Expression level.
The protein densities differ for each focal adhesion, suggesting variations in the expression level. However, regarding only focal adhesions of a single cell (red dots in the integrin plot), reveals that the density variations are based on individual focal adhesions. Therefore, differences in the expression level per se are of no decisive consequence for the further analysis. Instead, each adhesion site must be considered individually.

ent molecular organization of single focal adhesion sites. To answer this, all densities derived from several focal adhesions of a single cell were visualized.

In Figure 6.5, red spots are highlighted in the density variations of $\beta 3$-integrin and represent individual adhesion sites of only one single cell. This means, all adhesion sites are formed by a similar amount of endogenous and overexpressed $\beta 3$-integrin since they are all derived from the same cell, however the density variations are large. Obviously, these variations cannot be caused by the cellular expression level; instead focal adhesion sites themselves vary in their composition. With this observation the merging of all localizations together in order to obtain reliable structural information, as was done in section 6.1.2, seems questionable. Instead, single focal adhesions could be examined to exhibit their nano-organization.

6.1.4 Nano-polarity of single focal adhesions

It is well established, that focal adhesion sites arise in a polar manner [20, 157, 61], induced by retrograde actin flow and actomyosin contractility. The polarity could influence the molecular composition not only on the micro-scale but also on the nano-scale. Regarding this, each focal adhesion could

6.1. SPATIAL ORGANIZATION

have an intrinsic polarity, thus showing a different molecular architecture in the front (towards the cell edge) and in the back (towards the cell center) part of the adhesion site.

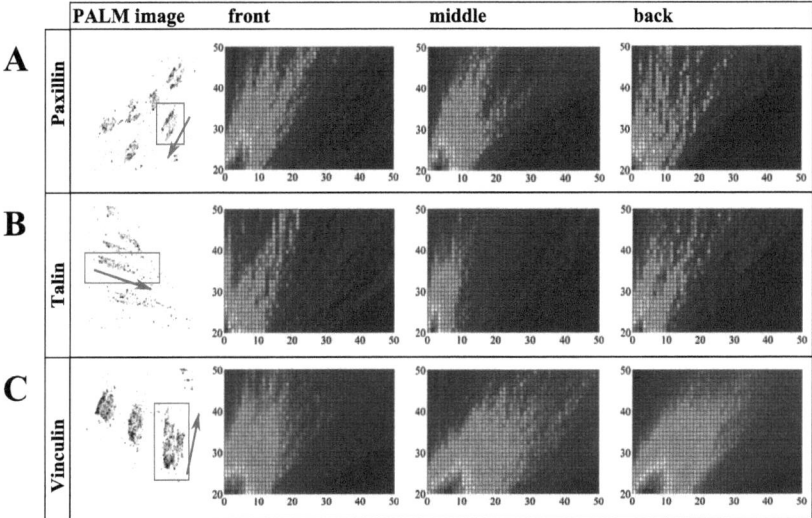

Figure 6.6: Investigation of an intrinsic polarity of focal adhesion sites.
The particular focal adhesions are marked with a red box and were observed in terms of their nano-organization. The arrow points in direction of cell migration (front). The degree distributions show different patterns, depending on the area within a single focal adhesion. However, there is no universal polar organization but an individual one for each focal adhesion. This suggests, that all adhesion sites have individual nano-organized structures.

In order to investigate a potential polarity in the nano-structure, single focal adhesion sites were separated into front, middle and back part, and analyzed to obtain the particular degree distributions. The resulting degree distributions exhibit a surprisingly modified picture. Some focal adhesions did not show significant alterations, while others reveal striking differences. Some example degree distributions are illustrated in Figure 6.6. The average peak value and its distribution varies according to the observed area individually. The neighborhood of Paxillin (Figure 6.6A), for example, exhibits a significant change in the back part of the focal adhesion. The degree distribution peaks at below 1 neighbor, whereas middle and front part have peak values for 3-5 neighbors. It is worth noting, that such results are not predictable by observing the corresponding PALM image. In contrast to Paxillin, Talin exhibits the lowest peak value in the front region, with increasing tendency towards the back part (Figure 6.6B). The last example degree distribution in Figure 6.6C is represented by Vinculin, which does not reveal large intrinsic alterations in terms of its neighborhood distribution between front, middle and back area.

These examples demonstrate a large variation of molecular organizations in a single focal adhesion. Each focal adhesion reveals individual alterations, which are not comparable to other focal adhesions. The polarity of adhesion sites is not reflected in a molecular density gradient. Even two focal adhesions, which are spatially close to each other, do not share the same characteristics in their degree distribution (not shown). This observation leads to the assumption that focal adhesions do not have a universally valid polar nano-organization; however they do contain spatial differences in their nano-architecture.

The output of the polarity check in Figure 6.6 proves that cell-matrix proteins cluster in different compositions, without apparent pattern. Hence, for a reliable information about clustering behavior of focal adhesion proteins, it is not advisable to pool several raw datasets together. Even the merged cluster information within a single focal adhesion contains only an average value, which does not represent the real situation.

6.2 Spatial organization in single focal adhesions

The intrinsic molecular organization of single focal adhesions differs significantly, without revealing a general polarity. The proximity information, obtained by degree distribution, further contains the information about the lateral xy-coordinates. Accordingly, the lateral positions can be plotted, including the proximity information of the degree distribution by color. By doing so, the mandatory radius cannot serve as a variable parameter anymore and must be determined.

In Figure 6.7 the spatial distribution of Paxillin inside a focal adhesion is depicted as a proximity plot. Strikingly, a single focal adhesion consists of areas with different protein densities, culminating in high density areas, as is highlighted in red. This confirms the assumption of section 6.1.2, after which the observed degree distributions are the results of rather large clusters. A super-resolution observation like this is not new, but was not further investigated so far [133].

6.3. PROTEIN NANO-CLUSTERS IN ADHESION SITES

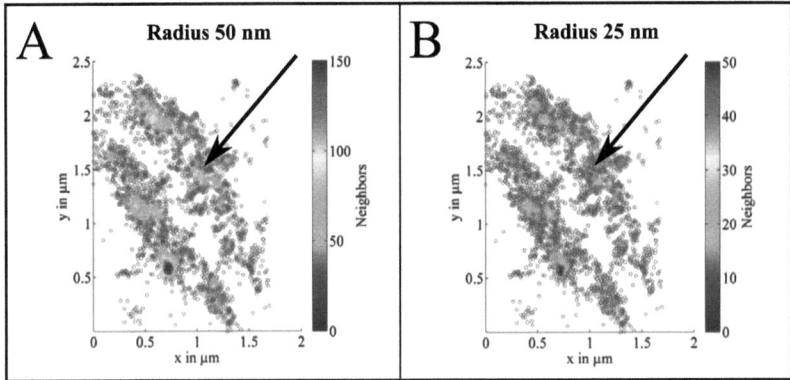

Figure 6.7: Lateral visualization of the degree distribution of Paxillin. Each localization is colored according to the number of its neighboring molecules in a radius of 50 nm (A) and 25 nm (B), respectively. Dense domains can be recognized in both plots. Plot B shows more details and furthermore detects smaller clustered domains (arrow).

The applied degree distribution in Figure 6.7B used a radius of 25 nm, which lies above the achievable resolution limit. In Figure 6.7A a radius of 50 nm was applied, which exhibits generally the same density distribution as Figure 6.7B, when the color bar is adjusted accordingly. However, clusters were more distinct with a smaller radius, visible in a more confined shape. Furthermore, rather small dense areas could not be traced with a radius of 50 nm, as indicated by the arrow. Consequently, a radius of 25 nm was generally applied to all reconstructed PALM-images, leading to a spatially resolved degree distribution of all observed proteins which can be further analyzed.

6.3 Protein nano-clusters in adhesion sites

Strikingly, all investigated cell-matrix proteins exhibit randomly distributed clustered domains within focal adhesions (not shown), matching the results from section 6.1.4, where a random distribution was predicted.

It was already shown that the nanometric organization is independent of polarity (cf. section 6.1.4). However, the cellular localization of focal adhesions could influence their structure. Distributions of focal adhesions in the cellular edge can be compared to focal adhesions localized in the cell body. It is likely that differences in density distributions will be observed, as focal adhesions in the leading edge serve different purposes than the ones in the cell body. Adhesion sites in the cellular edge do not only function as an anchoring tool, but also as environmental sensors [61], directly involved in cell signaling [159]. The edge contains the early lamella, and represents the area of focal complexes and focal adhesions. Accordingly, the maturation state ranges from focal complex to all maturation stages of a focal adhesion. In the cell body, mainly mature focal adhesions and fibrillar adhesions are present [77], whose main task is cellular anchorage. Furthermore, the edge regions are exposed to significantly stronger traction force (cf. chapter 1.4). Alterations in the molecular organization could be modulated by force and should be distinguishable when comparing the molecular densities of cellular edge and body.

The influence of force can further be studied by using force inhibiting drugs, such as Y-27632 (Y27) which prevents cellular traction force and causes the creeping degradation of actin bundles and focal adhesions, as it inhibits ROCK (Rho-associated kinase), and Cytochalasin D (CytD) which prevents F-actin polymerization by masking the plus-end, and thus induces F-actin degradation. The proximity distribution of all adhesion proteins was also studied after force inhibition. It was expected that force inhibition would produce a condition, similar to membrane localization outside of adhesion sites. Consequently such areas were selected in untreated cells. The resulting proximity distributions can serve as a reference.

6.3.1 Molecular density degree in adhesion sites

In order to distinguish the amount of dense versus less dense areas, a threshold for the maximal number of neighbors was applied to each proximity plot.

In section 6.1.3 it was demonstrated, that the difference in density is not derived from expression level variations, but from individual alterations among focal adhesions. All cell-matrix proteins reveal density differences in every single focal adhesion. Dense domains in generally less dense adhesion sites exhibit less neighbors than dense domains in dense adhesion sites. The focus here lies on dense domains in general, which includes also dense domains in less dense adhesion sites. In order to reach a statistical output, an individual threshold for each focal adhesion was applied by defining the maximal number of neighbors.

The results showed that the density differences in general were not that big. For TMD, Actin and α-actinin the lowest threshold, defining a dense domain, ranged between 45-55 neighbors. For all other proteins, a dense domain was defined by more than 40-45 neighbors. One Zyxin-transfected cell had generally very dense adhesion sites and therefore a threshold of

6.3. PROTEIN NANO-CLUSTERS IN ADHESION SITES 63

more than 60-65 neighboring molecules was set. A threshold of 50 neighbors was universally applied to proximity plots derived from force inhibited cells. After individual threshold adjustment of each proximity plot, the molecules in dense domains were indicated in red, and the molecules in sparse domains (under 20-30 neighbors) in blue. Molecules in intermediate domains were colored in green. The fraction of molecules in dense, intermediate and sparse domains were plotted in a bar diagram. Due to the individual adjustments, all percentage distribution of cellular edge, body or membrane, respectively, were merged in a single bar diagram, represented in Figure 6.8. The same procedure was applied to distributions of force inhibited cells.

It is worth mentioning, that the percentage distributions cannot be connected with the degree distribution from section 6.1.2. Formerly grouped proteins like Talin, Kindlin2 and Src do not exhibit similarities in their density distribution after manual threshold adjustment.

Cellular localization of adhesion sites

The percentage bar plots reveal no cellular localization dependency for protein aggregations in focal adhesions. Both, cell body and cell edge adhesion sites, show basically the same distribution for all cell-matrix adhesion proteins. A tendency in the direction of enhanced cell body localization of dense domains can be interpreted for Src, Actin and Zyxin. Furthermore, the cell body focal adhesions formed by over-expressed $\beta 3$-Integrin and FAK contain slightly more molecules in close proximity. However, considering the error bars, the amount of dense domains seems to be equal for cell body and cell edge for all proteins.

Cell-matrix proteins outside adhesion sites: An interesting observation can be derived from the reference showing the distribution of adhesion proteins outside of focal adhesions. Here, the same threshold was applied as for adhesion regions (40-60 neighbors). While there is nearly no dense membrane clustering for FAK, Talin and Actin, a fair amount of dense areas can be observed for Paxillin and α-actinin. Paxillin has many potential protein binding sites, which could cause an involvement in other unknown membrane processes. α-actinin contains a PIP2 binding site, which allows membrane localization. In addition, $\beta 3$-integrin has a tendency to form clusters outside of focal adhesions. This is not surprising since many transmembrane proteins exhibit a tendency to form clusters, as can be observed for TMD, as well.

A comparably large fraction of Vinculin, Src and Zyxin seems to cluster already in non-adhesive membrane regions. Src is also involved in other biological processes than cellular adhesions [6], and is generally not strongly

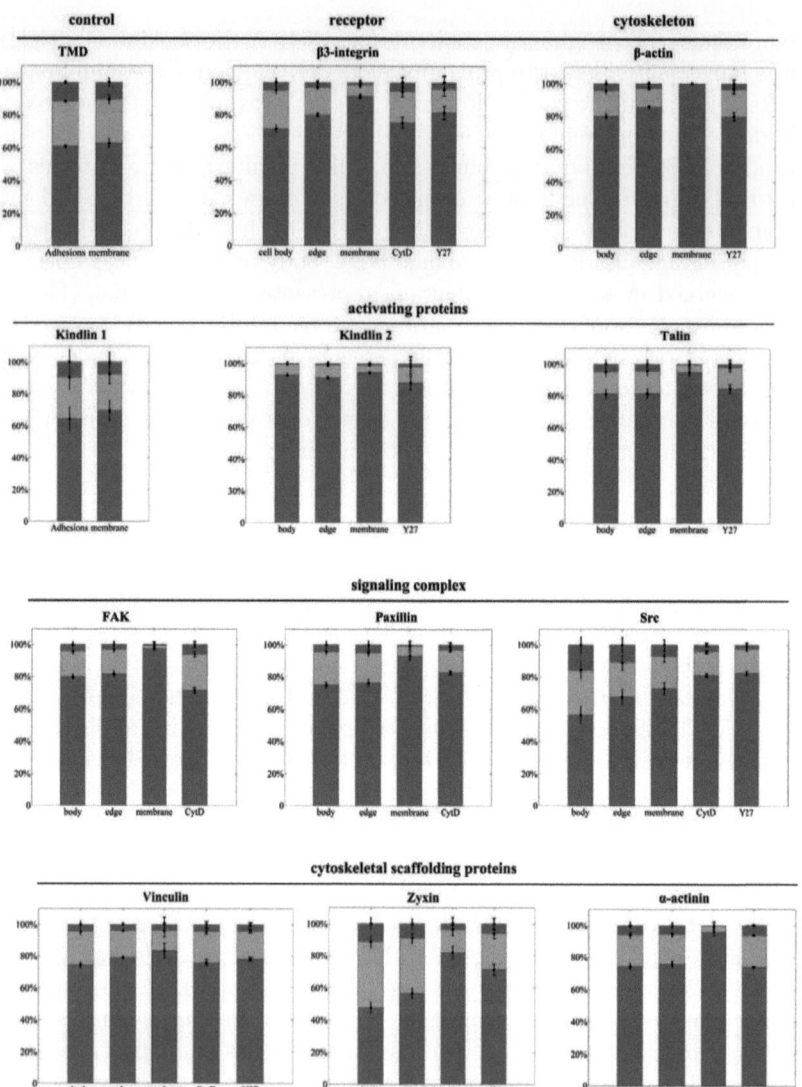

Figure 6.8: Percentage distribution of densities.
The percentage of molecules accumulating in high (red), medium (green) and low (blue) densities is illustrated. The amount of molecules in dense areas does not vary between cell body and cell edge, when the error bars are taken into account. Force inhibition by Y27 or CytD does not influence the molecular density. (Error bars shown are the standard error of the mean.)

6.3. PROTEIN NANO-CLUSTERS IN ADHESION SITES 65

localized in focal adhesions. Therefore, the enhanced membrane clustering could correspond to interactions in other signal transduction pathways, irrelevant for adhesion sites. Instead, Zyxin and Vinculin are known exclusively as adhesion proteins, which makes their membrane localization very interesting by itself. It could correspond to the unusual formation of nascent adhesions within the cell body, since Vinculin localization appears in the first steps of adhesion formation (cf. chapter 1.3). Zyxin instead, becomes only recruited after actomyosin activity, which excludes a localization in nascent adhesions. The high density outside of focal adhesions could also represent a pre-formed membrane complex or a membrane-standing left over of such proteins. However, this will not be further investigated in this work.

Controls for random spatial distribution: The investigation of the controls TMD and Kindlin1 exhibit interesting information as well. Since both proteins do not localize in focal adhesions, an additional FAK-immunostaining (cf. Chapter 4.3.2) was done to distinguish between adhesion and membrane area. There are no differences in density distribution between both areas, corresponding to their random distribution observed in section 6.1. However, a striking observation is represented by the large amount of dense domains in Kindlin1 and TMD control. Even though, the threshold setting was in the same range as for most other proteins, the number of highly dense and medium dense clusters is comparatively large. This speaks for a quite reasonable clustering affinity of randomly distributed proteins. It seems that proteins, involved in large protein assemblies like adhesion sites, do cluster even less than unspecific proteins. However, the induction of such clusters can be caused by the fixation treatment, as success and efficiency of fixation is still questionable [139].

Force dependency of dense domain formation

Figure 6.8 contains not only information about the density differences in terms of cellular localization of the focal adhesions. It can also be used to investigate the connection of dense areas to traction force, as it was mentioned that actomyosin contractility differs significantly between early and late adhesions.

The equal distribution of dense areas in cell edge and body for all proteins, observable in Figure 6.8 contradicts force dependency of protein aggregations. The presented data contains only focal adhesions in all maturation states, but no focal complexes or nascent adhesions (cf. sections 1.3 and 1.4 of Chapter 1), which could possibly exhibit a different picture, as force could influence the initial formation.

The effect of traction force inhibition is illustrated in the percentage bar diagrams derived from Y 27 and CytD incubated cells. Here, a distinction of different focal adhesion sites could not be made, as most adhesion sites disappeared completely, resulting in a merged area selection of former adhesion sites and membrane. Interestingly, the proximity plots of force inhibited proteins rather resemble the proximity plots of the adhesion sites than the ones of the membrane control. This stands for the surprising maintenance of adhesion proteins at the membrane, instead of being released into the cytosol. After drug treatment, no distinct focal adhesions could be observed by eye. However, on a nanometric level, fragments of the basic structure still remain at the membrane. Since force inhibition is a reversible process in living cells, it was observed that recovering adhesion sites assemble at similar positions as before the drug incubation (unpublished observation). Therefore, these dense areas, remaining after drug incubation, could function as the foundation of new adhesion sites.

Generally, these results support a force-independent formation and maintenance of dense domains. However, a potential function as modulator cannot be excluded, considering these results alone.

6.3.2 Cluster analysis

The proximity plots and their corresponding percentage bar diagrams already give some information about general distributions, but no information can be derived about the size and the amount of clusters. This might differ between edge and body due to cluster density variations, which cannot be reflected by the bar diagrams.

Dense domains were analyzed according to size and density using DB-SCAN cluster analysis and a complementary elliptical fit (cf. Chapter 5.1.2). The results derived by cluster analysis are listed in Table 6.2.

Most proteins show a mean cluster size of 50-65 nm and an average scattering of about 3 clusters/μm^2, regardless of their localization inside the cell.

The signaling proteins FAK and Src exhibit a slightly higher cluster size of 75-90 nm, whereas Src clusters must be much denser, according to the percentage distribution of Figure 6.8. Paxillin, which is known to interact tightly with FAK and Src, reveals larger clusters of about 100 nm size. Paxillin's average number of clusters per μm^2 is also a bit higher than the average number of most other cell-matrix proteins. However, according to Figure 6.8, the chosen threshold seems to fit since the density fractions are comparable to the ones of most other proteins. This alterations are probably linked to the focal adhesions selected for analysis, but could also indicate a difference

6.3. PROTEIN NANO-CLUSTERS IN ADHESION SITES

PROTEIN	Localization	AVG Diameter	StdDev	AVG No/µm²
Actin	edge	49	44	1.5
	body	51	41	2
	Y27	54	37	2
α-Actinin	edge	50	47	3
	body	53	35	3
	CytD	62	68	2
β3-Integrin	edge	56	45	3.5
	body	54	37	2
	Y27	25	15	4
	CytD	37	31	3
FAK	edge	76	40	2.5
	body	90	49	2.5
	Y27	67	29	4
Kindlin2	edge	25	4	1
	body	27	16	2
	Y27	45	42	1
Paxillin	edge	100	38	4
	body	107	70	4.5
	CytD	172	96	1.5
Src	edge	89	109	2.5
	body	76	85	3.5
	CytD	75	77	1.5
	Y27	57	61	1.5
Talin	edge	54	45	2
	body	44	41	2
	Y27	64	30	1.5
Vinculin	edge	53	45	2.5
	body	50	43	3.5
	Y27	58	62	2
	CytD	65	57	2.5
Zyxin	edge	260	116	7.5
	body	272	152	11
	Y27	199	92	3
Kindlin1	cell	73	59	3.5
TMD	cell	47	33	3.5

Table 6.2: Cluster analysis of dense domains inside focal adhesions.
Most clusters span a diameter of less than 100 nm. However, the variations of cluster size are rather big resulting in large standard deviations. A difference between cellular localization or upon force inhibition cannot be predicted. The average number of clusters per µm² shows also a similar distribution for most adhesion proteins.

in the localization of Paxillin. Paxillin plays a role, not only in the signaling complex, but it functions also as an important scaffolding protein which could explain larger Paxillin aggregates in focal adhesions.

The smallest clusters of under 30 nm are detected for Kindlin2, which shows also a very small percentage of dense areas in Figure 6.8. This indicates an incorrect threshold, which is further supported by the average number of clusters/µm^2 . Here, only a comparatively low amount of only 1 or 2 was calculated. An additional threshold check revealed an adequate choice, though. The localization of Kindlin2 in adhesion sites is rather loose, which could be accompanied by less aggregation. Furthermore, the loose localization complicates a proper threshold setting and therefore can still be a critical factor.

Zyxin in contrast, shows by comparison the biggest clusters of more than 250 nm in diameter. Furthermore, the average number of clusters is remarkably high, which corresponds to the density percentage in Figure 6.8. Zyxin revealed many dense areas of undefined shape, which were often connected with each other. The cluster parameters are derived from an elliptical fit, which could induce problems with poorly defined cluster shapes. Therefore, the cluster results of Zyxin should be treated with caution.

Kindlin1 and TMD as non-specific localizing proteins exhibit comparable cluster sizes and distributions. However, as Figure 6.8 shows, the percentage of dense areas is clearly higher than in adhesion proteins which implies a formation of denser clusters.

The cluster analysis for force inhibited (Y27, CytD) samples reveals interesting information. According to Figure 6.8, the number of molecules forming dense domains remain mainly the same compared to focal adhesions. Table 6.2 reveals that the size of a dense domain seems to increase upon force inhibition. The average clusters per µm^2 support this impression by a lower number compared to untreated samples. Therefore, the dense domains observed in focal adhesions, seem to rearrange into bigger, but not denser clusters. The only significant outlier is represented by β3-integrin, which exhibits cluster shrinkage upon force inhibition. Such a result is difficult to place, because especially β3-integrin was expected to be less affected than all other cell-matrix proteins. As a receptor, β3-integrin is anchored to the ECM which implies an additional stabilization of its position. The fact, that all other adhesion proteins remain on the membrane implies further that β3-integrin serves as foundation for their accumulation. Both cannot be confirmed by the calculated cluster parameters of Table 6.2. The density percentage diagram in Figure 6.8 would predict another result, as well. Therefore, the cluster shrinkage of β3-integrin upon force inhibition remains puzzling.

6.3. PROTEIN NANO-CLUSTERS IN ADHESION SITES

Difficulties with cluster definition: Regarding the standard deviation, the interpretation above becomes rather relative. For most proteins, the variation is in the same range as the diameter itself. This is mainly caused by the large differences in cluster size itself.

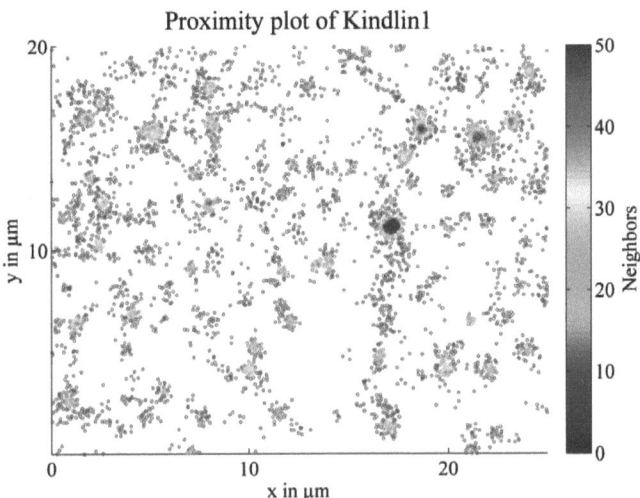

Figure 6.9: Proximity plot of Kindlin1. A large membrane region is illustrated, which exhibits Kindlin1 aggregations in all densities and sizes. Furthermore, most dense areas are not round but elliptic or of undefinable shape. This example mirrors the observations of all other investigated adhesion proteins, which show similar variations of dense domains.

Figure 6.9 shows an example proximity plot of Kindlin1. It exhibits several dense domains, which largely differ in their density, as can be distinguished by color. Therefore, already the choice of the right threshold is not trivial and continues with a harsh cluster analysis. When a dataset contains big and small clustered regions of diverse densities, the software applies a threshold corresponding only to the regions of high density. As a result, dense clusters will mask the less dense clusters, which is entailed by the integration of only a small part of less dense clusters. This affects the average cluster size as well as the standard deviation. Furthermore, the clusters are not uniformly round, but have elliptical shapes up to undefined formations, which can cause fitting problems. The presented analysis in Figure 6.8 and

Table 6.2 can give evidence for a possible clustering of focal adhesion proteins, but does not reflect the real observations in a reasonable manner.

An acceptable conclusion of the presented cluster interpretation would be that the observed clusters alter remarkably in size, position and density even in a single focal adhesion. However, they do not exceed a size of 150 nm. Besides, the clusters appear in all maturation states of focal adhesions and do not differ significantly among the cell-matrix proteins. The mentioned differences observed for Kindlin2 and Zyxin as well as for the force inhibited β3-integrin could be a result of inaccurate thresholding or cluster definition and fitting.

The results of force inhibited samples reveal not only the persistence of nanometric clusters, but suggest even an enlargement compared to adhesion structures. However, it is not traceable if these structures are localized in former focal adhesions or if they are equally distributed in all membrane areas.

6.4 Speculations concerning protein accumulations

According to the conclusion of section 6.3.2, the protein aggregates within focal adhesions most likely show a similar size and distribution across all investigated proteins. Therefore, it can be speculated that all involved proteins accumulate together in these dense clusters. Protein aggregates could represent a potential center for signal transduction, or serve as a foundation for strong traction force. In sections 6.1.4 and 6.2 it was demonstrated that there is no visible intrinsic pattern, nor a general polarity. Instead, highly dense domains scatter randomly in the entire adhesion site. Furthermore, their appearance does not correlate with the localization or size of focal adhesions.

The PALM results suggest a force-independent formation of these dense domains. However, the initial application of traction force could induce protein aggregation by recruitment of supporting scaffolding proteins. No nascent adhesions, which are generally force-independent, were observed in this chapter to study this hypothesis. However, force inhibition did not negatively affect dense areas. Instead, force inhibition led to a slight increase of cluster size for most cell-matrix proteins.

Protein aggregations distribute equally in all focal adhesion maturation states, starting from young adhesion sites in the early lamella. This indicates an early formation of dense domains and a late disassembly. Once formed, dense domains could persist until turn-over in the focal adhesions. Therefore,

6.4. SPECULATIONS CONCERNING AGGREGATES

dense areas could occur via diffusion-driven formation, most probably based on integrin receptors.

The steps of adhesion formation are still not completely understood, but receptor-ligand binding is mandatory and induces further integrin clustering. Since a recruitment mechanism is very unlikely for transmembrane proteins, integrin clustering emerges, most probably, by randomly diffusing integrins, hitting immobile receptors. An eventual integrin-integrin affinity would induce deceleration and could increase the probability for integrin receptors of getting trapped by ligand or Talin. Such diffusion-driven integrin clusters recruit cytosolic proteins, which in turn form protein clusters in response to their binding to the dense integrin-tails. In this case, a directed functionality is questionable, though a signaling center is possible.

Both hypotheses cannot be verified by PALM experiments, because fixed cell experiments provide no information about protein dynamics. However live cell experiments can be performed to study the dynamic changes of clustered areas, as was done in the following Chapters.

Chapter 7

Single particle tracking in living cells

In order to further investigate the dense domains discussed in Chapter 6, single particle tracking was performed to monitor the dynamic behavior of the transmembrane receptor β3-integrin, tagged with mEos2. Temporal alterations of the diffusion constant were observed when inducing the formation and degradation of focal adhesions. All other proteins are recruited from the cytosol, which prevents the calculation of diffusion constants. Instead, a spot-activation experiment was performed, to unravel whether the observed protein aggregations are targets of enhanced adhesion protein recruitment.

7.1 Mobility changes of β3-integrin upon force inhibition

In Chapter 6.2 it was shown that force inhibiting drugs do not significantly influence the density distribution of cell-matrix proteins. Furthermore, protein aggregations persist even after force inhibition, suggesting force-independency. However, these PALM experiments were performed upon high cellular stress since the time of drug incubation was comparably long, followed directly by fixation. After this kind of treatment, the formation of artifacts is possible. Hence, such results obtained by PALM should be validated with additional live cell experiments.

If traction force induces protein aggregation, force alterations should affect the size and number of dense domains. The PALM results contradict this, by showing equal amounts of dense domains in all maturation states of focal adhesions (cf. Figure 6.8). In living cells, this result can be verified by visualizing dynamic alterations upon force inhibition. If traction force functions as a major inducer of dense domains, subtraction of force should even lead to the rapid disassembly of highly clustered domains. Further, the lack of traction force should lead to an equally dense distribution for all remaining adhesion fragments.

7.1.1 β3-integrin dynamics influenced by force inhibition

Figure 7.1B shows the disassembly of the force-dependent adhesion protein Zyxin upon Y27 incubation in HeLa cells. Zyxin was co-transfected as a reference for successful force inhibition, as the creeping disappearance of Zyxin correlates with the loss of force in adhesion sites. β3-integrin is also influenced by Y27 but with less obvious effect, as illustrated in Figure 7.1A. This effect can be better demonstrated by changes in the mobility of β3-integrin. Force inhibition supports the release of former immobile integrin receptors from adhesion sites, indicated by increased diffusion constants, as shown in Figure 7.1C.

The effect of Y27 on each cell can vary upon incubation time (5-30 min), depending on the quality of Y27 as well as on the individual reaction of the cell upon drug incubation. Furthermore, the recovery rate after Y27 wash out differs among cells and can take 20 to 60 min. Due to these efficiency variations, the averaged diagram in Figure 7.1C (left) provides only information about initial drug incubation (Y27) and the subsequent time point for drug wash out (W/O) directly after complete Zyxin loss. The average diffusion constant of each cell can yield different values, too. This is mainly caused by the number of focal adhesions. In a cell with many adhesion sites, a large fraction of β3-integrin is immobile, while cells with less focal adhesion sites have a larger fraction of β3-integrin in the unbound, mobile phase. As a consequence, all values were normalized to the initial diffusion before drug incubation. Despite considering such disturbing factors, the observed Y27 effect in the normalized diagram does show little effect and large error bars. Therefore, the diagram of a single experiment with apparent changes is shown in Figure 7.1C (right), which in addition provides an insight into the average diffusion constants.

When Y27 is removed, the cell can recover by forming new focal adhesion sites, which is indicated by a drop of the average diffusion constant in Figure 7.1C. As the amount of newly formed adhesion sites can be different from the amount before drug treatment, no definite conclusion can be given about the completeness of recovery. A cell contains two major integrin receptor states. The active ligand-bound state in focal adhesions and the

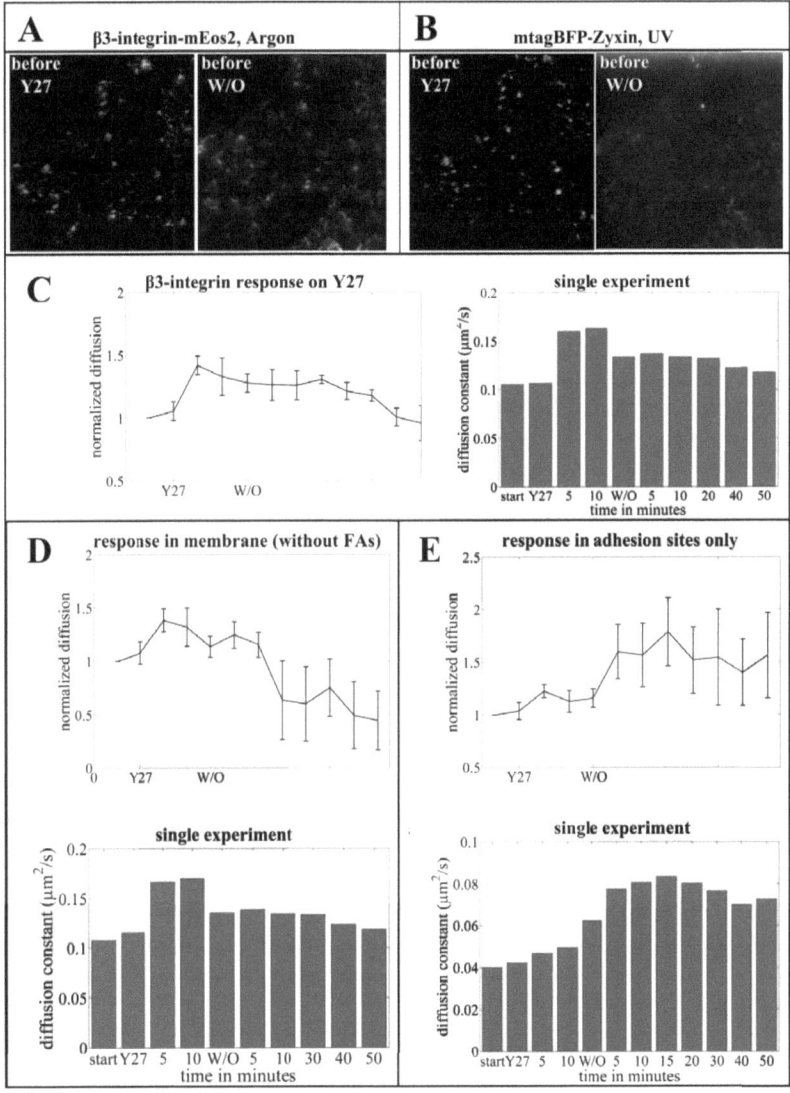

Figure 7.1: Effect of Y27 incubation.
Y27 induces focal adhesion disassembly. Subtle changes are observable for β3-integrin (A), while Zyxin disassembles completely (B). The average diffusion constant of β3-integrin increases upon Y27 incubation and drops again after wash out, as can be observed in the normalized diagram as well as in the single experiment (C-E).
The start-value represents β3-integrin diffusion before Y27 treatment, Y27 indicates the time point of Y27 addition and W/O stands for the timepoint of wash-out. Error bars shown are the standard error of the mean.

inactive state, diffusing in the membrane outside of focal adhesions. Therefore, it should be distinguished between focal adhesion area and remaining membrane area, to observe the temporal mobility changes of $\beta 3$-integrin. In order to remove the trajectories inside the focal adhesions, the wide field localization of mtagBFP-Zyxin directly after addition of Y27 was used as reference image for adhesion sites. A mask of this image was generated and then applied to the $\beta 3$-integrin trajectories. As Zyxin is only recruited to focal adhesions being exposed to actomyosin contractility, the choice as a reference protein seems valid.

Mobility in the membrane outside of focal adhesions: In Figure 7.1D only membrane regions without focal adhesion sites are considered. Interestingly, the average diffusion constant of $\beta 3$-integrin increases after Y27 incubation. This result is surprising, as freely diffusing integrin receptors should not be affected by force inhibition. Nascent adhesion complexes, which contain immobile integrin receptors, are not yet connected to the cytoskeleton and cannot be impaired by Y27. Therefore, the observed mobility increase is puzzling and might be an artifact of the mask application. It is conceivable, that small focal contact sites were not included, because the recruitment of transfected Zyxin was not sufficient to create a visible localization in widefield. In turn, the trajectories occurring in these regions are also included in the diagram of remaining membrane. As both, single experiment diagram and the normalized diagram reveal a visible mobility increase of $\beta 3$-integrin upon Y27 addition, an inefficient mask-selection of focal adhesions seems plausible. Apart from that, a clear decrease of integrin receptor mobility is visible, as soon as Y27 was washed out. This effect can be attributed to the formation of new focal complexes, followed by their maturation into focal adhesions. The recovery of focal adhesion sites was often not very efficient, which is indicated by the large error bars after wash out in the normalized diffusion in Figure 7.1D. In case of poor recovery, $\beta 3$-integrin does not immobilize to form new adhesion sites and maintains a rather high average diffusion constant.

Mobility in adhesion sites: The alterations exclusively inside focal adhesions are shown in Figure 7.1E, based on the same mtagBFP-Zyxin image as a reference mask. Interestingly, the diffusion constant increases just very slightly during Y27 incubation. Only after drug wash out, the diffusion constants rise significantly. A possible reason can be found in the anchorage of integrin receptors. The immobility of integrins is not only caused by the intracellular connection to the actin cytoskeleton, but also by its binding to

7.1. MOBILITY CHANGES UPON FORCE INHIBITION 77

the extracellular matrix. Accordingly, the Y27 induced intracellular integrin release does not automatically cause a ligand detachment. Hence, most integrin receptors remain at their static position and show only minor diffusion changes. The supportive Figure 7.2 shows the average diffusion constants of the mobile ($>0.2\,\mu m^2/s$), immobile ($<0.1\,\mu m^2/s$) and intermediate fraction and reveals observable changes. The immobile fraction decreases, while intermediate and mobile phase increase upon Y27 incubation. However, the largest fraction remains immobile and drops only by about 10% compared to the reference diffusion before Y27 incubation (start). As the effect is rather small, other turn-over mechanisms might take over. β3-integrin could disappear by endocytosis or simply by a lack of efficient β3-integrin exchange. Low exchange rates would not reduce the amount of integrins in the vanishing focal adhesion, but could induce the false impression of strong β3-integrin reduction, as bleached molecules will not be exchanged.

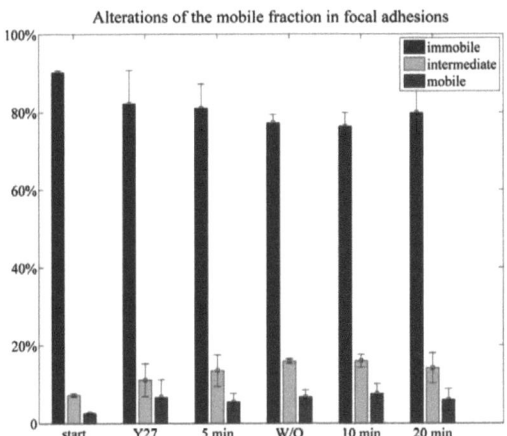

Figure 7.2: Mobility changes of β3-integrin in focal adhesions. Y27 affects the mobility of β3-integrin inside focal adhesion sites. The diffusion constant of the mobile and intermediate fraction increase directly after Y27 addition, while the immobile fraction decreases. However, the changes are not that big, as the immobile fraction falls by average 10% until drug removal. (The mobility classification corresponds to [117].)

When Y27 is removed, possibly the buffer flow breaks the binding between ligand and integrin and causes an increase in diffusion, as observable in Figure 7.1E. From the moment of wash out, β3-integrin's mobility seems to stay relatively constant. Still, noteworthy are the large error bars, as were seen also for the recovery of remaining membrane (Figure 7.1D). The same principle can be applied in regions of former adhesion sites. Some of the initial focal adhesion sites recovered (the diffusion constant dropped) while others disassembled completely or reassembled in new regions (diffusion constant reached plateau). A look at the single experiment in Figure 7.1E exhibits a maximum value at 15 min after wash out for the average diffusion constant. Then the mobility declines again, but does not reach the initial value. Here, focal adhesions did probably mainly disassemble but few recovered in the initial area.

7.2 Density alterations upon force inhibition

In section 7.1 it was demonstrated that alterations in the diffusion of β3-integrin receptors caused by Y27 incubation can be mapped by calculating the diffusion constants. Since the largest fraction of the β3-integrin receptor remains immobile in adhesion sites, the maintenance of dense domains and their dependency on force can be studied.

If dense domains are induced by traction force, an asymmetric disassembly of focal adhesion sites could be an indicator, since the release of strong pulling forces should lead to rapid vanishing of the dense domains. A protein cluster should disassemble as soon as the actomyosin contractility in this region stops. Therefore, the diffusion constants of β3-integrins inside a cluster should be similar in the short time interval of release. Furthermore, the force inhibition should induce a complete vanishing of dense domains and leave an equal distribution.

Trajectories of β3-integrin-mEos2 were detected during Y27 incubation in order to spot a potential loss of the dense domain structure. To further verify the spatial organization, the number of immobile neighbors in a radius of 150 nm was calculated according to Chapter 5.1.2. This radius corresponds to the largest protein clusters observed and analyzed in Chapter 6. In live cell experiments - as in PALM experiments - only single molecules were observed. Still, the resolution in a living cell experiment is always lower than in a PALM experiment, due to cell and protein dynamics, which allows the application of a rather large radius.

Figure 7.3 visualizes the neighborhood changes inside focal adhesions of two Y27 incubated cells. Both cells show a degree distribution peaking at around 5 to 10 connections. No significant changes in the number of neighbors can be detected upon progressing Y27 incubation, which reflects the

7.2. DENSITY ALTERATIONS UPON FORCE INHIBITION

maintenance of the initial average cluster distribution. This reveals that actomyosin contractility loss does not affect the cluster degree and does not change the dense domain structure per se. Only the removal of the drug causes the disassembly of the remaining structures. According to this observation, the general density of β3-integrins seems to remain mainly equal upon Y27 incubation, which corresponds to the small changes in the diffusion and mobility, observed in the Figures 7.1 and 7.2.

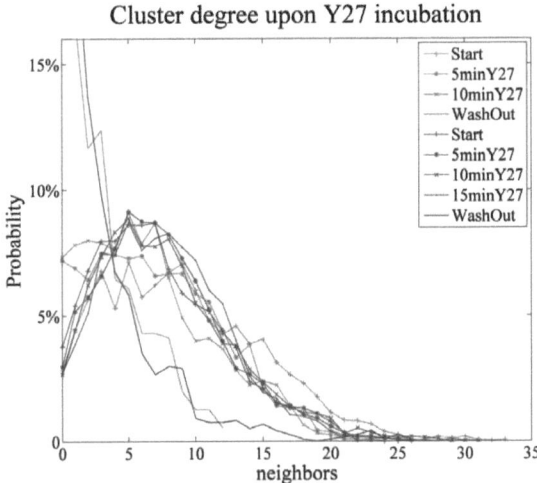

Figure 7.3: Alterations in the degree distribution during Y27 incubation.
The loss of traction force does not affect the average distribution of β3-integrin receptors in a radius of 150 nm, which indicates the maintenance of the preformed cluster structure. Only drug removal induces the disassembly of clusters.

7.2.1 Dynamic distribution of β3-integrin in adhesion sites

Force inhibition could affect only a subset of clustered areas, which will not be mirrored in the average degree distribution. Therefore, only some dense domains would disassemble quite rapidly, visible by higher individual diffusion constants in a nearby area.

An example focal adhesion is visualized in Figure 7.4, which distinguishes between mobile and immobile molecules [117]. Only the first localization of

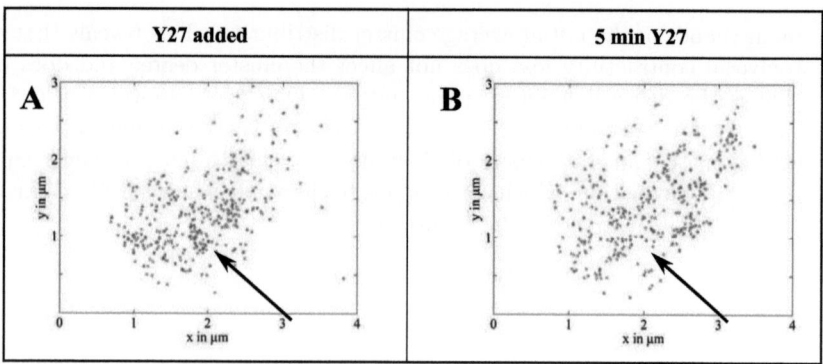

Figure 7.4: Distribution of the mobile (blue) and immobile (red) fraction of β3-integrin upon Y27 incubation.
The mobile fraction is equally distributed in the entire adhesion site (A and B), which does not support a sudden release of β3-integrin upon traction force loss. However, former dense domains seem to vanish (indicated with an arrow) while dense areas form in other areas.

each trajectory was plotted as a reference for its initial position. In Figure 7.4A some potential dense domains are visible by eye, which are mainly immobile. Still, some mobile β3-integrin molecules are scattered quite equally over the entire focal adhesion, including the rather dense areas. In Figure 7.4B, neither an accumulation of mobile molecules can be observed nor a gradient that would indicate an asymmetric turn-over. The distribution of mobile molecules in a focal adhesion contradicts traction force dependency. However, an argument for force-dependency can be found since the dense domains, visible at Y27 addition, are completely vanished after 5 min incubation time, as can be observed by holes in the middle of the focal adhesion (arrow). Such an observation could be caused by the bleaching of β3-integrin-mEos2 in this region. However, a dense domain contains at least 50 molecules (cf. Chapter 6.3), so it seems rather unlikely that all of them become activated and bleached during one single measuring cycle.

The indications for a possible force dependency of dense domains in Figure 7.4 must be further investigated. A distinction between mobile and immobile receptors might be not sensitive enough to spot significant mobility changes of β3-integrin.

Force inhibition induces the release of focal adhesion proteins, as soon as the cytoskeletal anchorage breaks, which is connected with a large increase of mobility, however, untraceable with SPT. Integrin receptors are additionally anchored by extracellular ligand binding, which keeps them mainly static.

7.2. DENSITY ALTERATIONS UPON FORCE INHIBITION

Still, the general stability should be reduced in the absence of a cytosolic anchor, which should be indicated by an increasing mobility of the former immobile fraction.

The alterations of the individual diffusion constants in a single focal adhesion, as well as the corresponding neighbor proximity in a 150 nm radius are plotted in Figure 7.5. The diffusion constant of each trajectory was calculated and plotted in Figure 7.5A, according to the localization coordinates of the first frame. The different diffusion constants can be distinguished by color. The colorbar was was set to a maximum of 0.1, reflecting the rather immobile state. However, this adaption also reveals a very equal mobility distribution during the whole experiment. Again, no accumulations of similar diffusion constants can be observed, indicating no sudden release of dense integrin receptor domains upon force relaxation. This result supports force-independent protein accumulations.

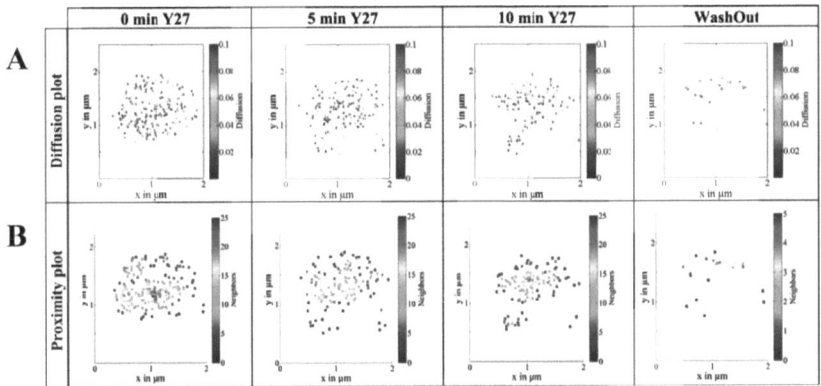

Figure 7.5: Spatial changes in diffusion and density.
The individual diffusion constants of $\beta 3$-integrin molecules are equally distributed (A). No gradient or clustered release of receptors is visible. The localization of dense areas seems to change upon Y27 incubation (B). Even after drug removal areas of receptors remain in close proximity.

The proximity plot of the same adhesion site in Figure 7.5B contains some surprising observations. The initial proximity plot reveals some dense domains indicated by color. Upon progressing focal adhesion loss, the general structure of the remaining focal adhesion also changes. A loss of $\beta 3$-integrin in the lower right part can be observed, involving also parts of the initial dense domains. However, the other initial dense areas are still observable after 5 min of Y27 incubation. Interestingly, at 10 min Y27 incubation the dense areas re-form or alter their shape and size, which allows the assumption that force could appear as a modulator of dense areas.

After wash out it was further observed, that some of the remaining $\beta 3$-integrin receptors seem to remain still in close proximity, as if some dense domains would survive the drug treatment. A similar observation was already obtained in Chapter 6.3.2, which suggested the maintenance of $\beta 3$-integrin in smaller clustered structures at the membrane after force inhibition. These remaining immobile receptors could cause the cellular re-adhesion at a similar position after drug recovery, which was observed several times (cf. section 7.1).

Asymmetric disassembly at 5 min Y27 and alterations in dense domains at 10 min Y27 support a force-dependency of dense domains. A plausible explanation for this phenomenon without considering force-influence could be the experimental procedure itself. First, only a small subset of molecules is activated to generate an image. This might result in an inaccuracy in localizing cluster centers, visualized as red spots in Figure 7.5. Second, fluorophore bleaching could have large impact. It was already mentioned that force inhibition might negatively influence the exchange rates of integrin. As integrin receptors show generally a low exchange rate of 2-4 minutes [15, 112], the measurement intervals of 5 min may be too short for recovery.

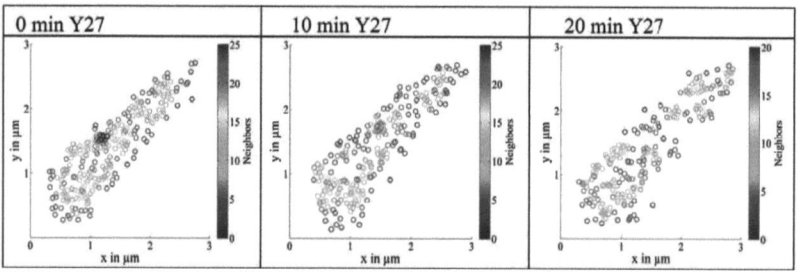

Figure 7.6: Density alterations upon Y27 incubation.
Areas of high density (dark red) do not disassemble or change position in the first 10 min. Smaller dense areas disassemble and are newly formed, or change position. After 20 min a faint dense area remains, but its position does not exactly match the position of the initial dense domain.

Therefore, the same experiment was performed, detecting more particles in a larger time interval of 10 min, which should diminish the measurement artifacts. Figure 7.6 demonstrates the result on the basis of a single focal adhesion, showing the same phenomenon as in Figure 7.5. Initially dense areas change its position within the first 10 minutes of Y27 incubation. Another

10 minutes later, one dense domain disappeared completely, while another one remained but revealed again fluctuations in its localization. Generally, the density distribution seems more equal, which could be simply caused by the release of integrins due to Y27 incubation, but could be also a sign for a force-dependent vanishing of dense areas. It is worth mentioning that the focal adhesion marker Zyxin disappeared between 10 and 20 min, which means that the distribution of Figure 7.6 (right) must be force-independently formed. However, as alterations happened also before, a modulation function of force is possible.

These observations are demonstrated only by means of three focal adhesion sites, but alterations in size and position of dense domains were observed universally for β3-integrin in most focal adhesions influenced by Y27.

7.3 Induction of the high-affinity state of integrin

The results obtained in section 7.1 cannot clearly define, whether traction force has an influence on dense domains in focal adhesions. Most results represent the general maintenance of dense domains during force inhibition, which supports force-independency. However, alterations in shape and position of integrin aggregations have been observed, placing force into the role of a modulator of dense domains. Hence, the initial formation of dense domains could also be completely force independent.

Another theory is based on the observations made in Chapter 6, in which randomly diffusing proteins have been shown to have a tendency to cluster. Accordingly, the initial formation of integrin clusters could be simply diffusion-driven and pursuing an intracellular recruitment of cell-matrix proteins, which then culminate in the formation of an adhesion site. The first ligand-bound immobile integrin receptor could induce the deceleration of freely diffusing integrin receptors by clashing. If the clashed receptor is also in an extended conformation, the mobility decrease could cause direct ligand binding. A slowly moving inactive receptor could be easier accessible for ligand or Talin, in terms of activation. Therefore, this cluster formation is further referred to diffusion-driven clustering. In order to investigate this hypothesis, outside-in activation of β3-integrin must be initiated. Manganese ions have been proven to be effective initiators. The divalent cation Mn^{2+} rapidly exchanges the Mg^{2+} ion in the MIDAS [130] and the Ca^{2+} ion in the SyMBS as well as the Ca^{2+} ion of the ADMIDAS [161], all present in the I-domain of the integrin receptor. Mn^{2+} ion coordination in the I-domain mimics ligand binding [30] and therefore induces the conformational change of the integrin receptor (cf. Figure 1.1). In its extended, high-affinity conformation, integrin exhibits its ligand binding site which becomes accessible by a ligand of the extracellular matrix.

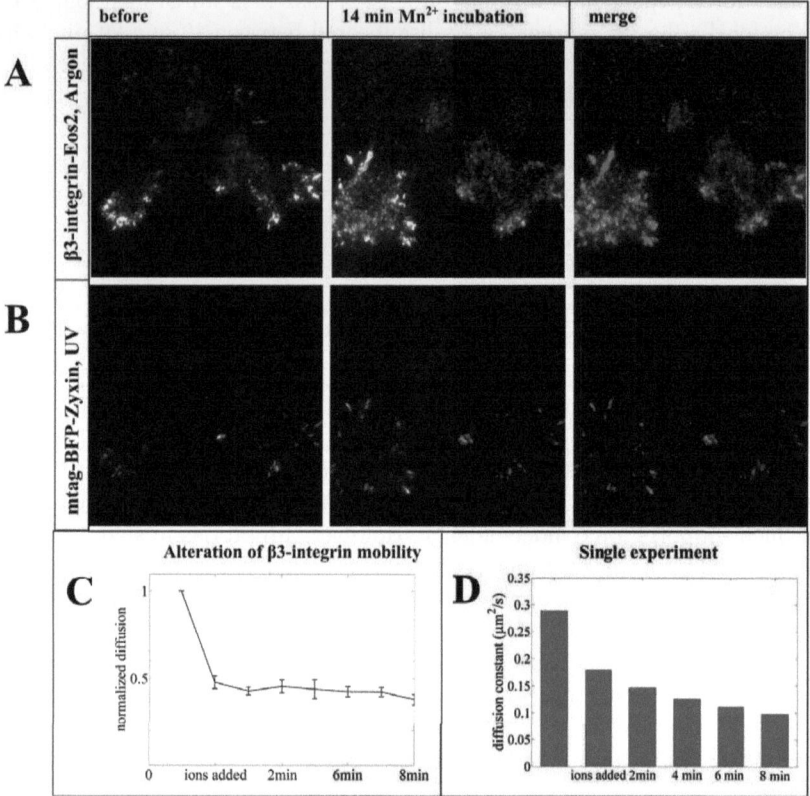

Figure 7.7: **Mn^{2+} ions induce clustering and cell spreading.**
Incubation with Mn^{2+} ion induces the high-affinity state of integrin receptors, which goes together with enhanced membrane clustering (A) and a sudden decrease in mobility (C, D). Zyxin is not affected by integrin clustering, since there is no enhanced membrane accumulation (B). Furthermore, an enhancement in cell spreading upon Mn^{2+} incubation is visible (A, B). Zyxin can function as a marker for the formation of new adhesion sites. (Error bars shown are the standard error of the mean.)

In Figure 7.7A the effect of Mn^{2+} ions on integrins is strikingly demonstrated. The fluorescence images clearly show an enhancement of the β3-

7.3. INDUCTION OF THE HIGH-AFFINITY STATE OF INTEGRIN

integrin membrane level, as well as the assembly in a clustered organization. Figure 7.7B shows the fluorescence images of Zyxin referred to Mn^{2+} ion incubation. The alteration of Zyxin localization can be attributed to cell migration rather than membrane recruitment due to Mn^{2+} ion incubation. Zyxin accumulates only in very distinct areas, whereas β3-integrin shows a quite unsystematic distribution. This reveals that integrin clustering does not automatically induce the recruitment of cytosolic cell-matrix proteins, resulting in the formation of new focal adhesion sites. So far, recruitment - directly connected to the incubation with Mn^{2+} ions - was only observed for the inside-out-activator Talin [36] and its supporter Kindlin2, as was shown in Figure 7.8. However, the recruitment of Talin does not automatically result in adhesion formation, even though Talin can induce inside-out-activation. Most likely, only the Calpain-cleaved Talin head-domain is recruited, whereas full-length Talin will be not attracted by Mn^{2+} ion induced integrin extension.

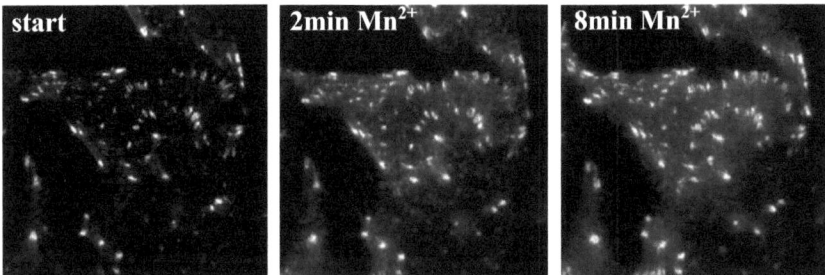

Figure 7.8: Kindlin2 response on Mn^{2+} ions.
Mn^{2+} incubation induces the recruitment of Kindlin2 to the cell membrane.

The induction of the high-affinity state causes an immediate decrease in mobility of nearly half the average diffusion constant, as shown in Figure 7.7C, corresponding to the literature [134, 117]. The normalized curve, as well as the example plot in Figure 7.7D clearly show that under Mn^{2+} containing buffer conditions the mobility of β3-integrin cannot recover. Instead, the initial drop represents already roughly the maximal drop, which implies a maximal degree of immobilization from the very beginning. This leads to the assumption that integrin receptors in the extended conformation become directly bound by a ligand of the ECM.

Regarding only the influence of Mn^{2+} ions on the alteration of integrin clusters outside adhesions, degree distribution was applied to immobile molecules and plotted in Figure 7.9. The focus here is on initial cluster formation based on the reduction of the velocity of β3-integrin, which excludes

86   CHAPTER 7. SINGLE PARTICLE TRACKING IN LIVING CELLS

the consideration of dense domains in adhesion sites. For degree distribution calculations a smaller radius of only 100 nm was applied, as this reveals already significant differences. A membrane reconstruction exhibits a clear enhancement of clustered areas, visualized by the colorbar (Figure 7.9A) and the diagram in Figure 7.9B, which shows an altered connectivity upon Mn^{2+} ion treatment. In this experiments, the number of observed unbound β3-integrin receptors (less than one connection) reduced to half after addition of Mn^{2+} ions.

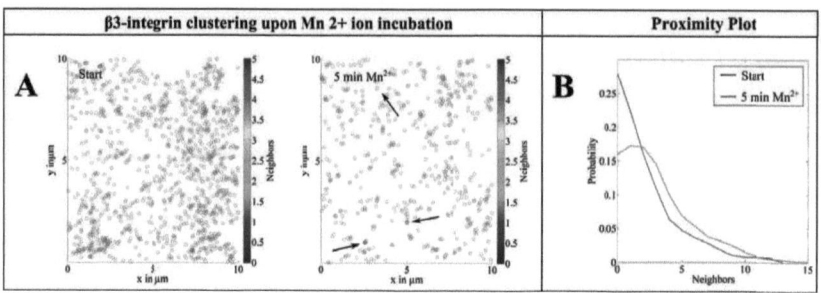

Figure 7.9: β3-integrin membrane clustering upon Mn^{2+} treatment.
β3-integrin shows enhanced clustering after the addition of Mn^{2+} ions, indicated by arrows (A). One red dot consists of at least 5 molecules in a radius of 100 nm. Also the degree distribution reveals a clear decrease of unbound receptors (B). The favored cluster composition seems to consist of 2-4 molecules after Mn^{2+} addition.

7.4 Formation of new focal adhesion sites

In section 7.3 it was shown that Mn^{2+} incubation reduces the velocity of β3-integrin and stimulates the accumulation of clusters. These clusters have no biological purpose and are modulated only by diffusion, as no cell response in terms of recruitment of multiple other cell-matrix proteins, leading to the formation of functional adhesion plaques, is visible. However, in rare cases, new adhesion sites appear during such an experiment. Since Mn^{2+} ion influence is limited only to receptors and first binding partners, the further recruitment of signaling and scaffolding proteins should be simply based on diffusion based integrin clustering. Since all integrin receptors are in the extended conformation, the adhesion formation should be induced by outside-in-activation. Adhesion site induction, caused by the inside-out-mechanism

7.4. FORMATION OF NEW FOCAL ADHESION SITES

(cf. Chapter 1.2.1) cannot be excluded, but is rather unlikely, since the large amount of pre-extended integrin receptors - due to Mn^{2+} ion incubation - should diminish the probability of inside-out-activation. The assembly of co-transfected mechano-sensitive Zyxin serves as indicator for the formation of functional focal adhesions. In Figure 7.7B, the merged fluorescence image also exhibits cellular spreading during the experiment which is a known effect of Mn^{2+} ions [93, 63]. As a result, the formation of new adhesion sites can be investigated already from the nascent state, followed by maturation to focal complexes and further to focal adhesions.

The temporal assembly of β3-integrin receptors in new focal adhesions is visualized in Figure 7.10. After 2 min, Mn^{2+} ions were added, and another 6 min later, most of the new formed adhesion sites started to grow. For many of these growth curves, an increasing slope can be observed after 10 or 14 min, respectively, which can be misinterpreted as an enhanced recruitment rate. Instead, it represents the forced break of the measurements, due to the limited memory storage of the computer, which required about 20 additional seconds. Apart of these artifacts, no further irregularities were observed in the slopes. Formation of adhesion sites seem to progress rather steadily, as the curves are nearly linear. Zyxin recruitment proves the application of actomyosin contractility upon focal complex formation and further adhesion maturation, but cannot be picked out from this graph. In other words, force does not lead to accelerated protein recruitment, which would induce a sudden slope increase. This fact further substantiates the hypothesis of an traction-force independent formation of dense domains.

A proximity reconstruction, showing the formation of new adhesion sites, is provided in Figure 7.11. The colorbar indicates the proximity of molecules calculated for 150 nm. The formation of dense domains can be followed from the initial detection point of immobile β3-integrin receptors until the end point of the measurement.

First, only few molecules immobilize, which already revealing an interesting pattern. In both examples (Figure 7.11A and B), some receptors immobilize in rather close proximity of 150 nm, whereas others are far away, which indicates that initial adhesion formation is most probably not based on a single immobile β3-integrin receptor. The receptors in rather close proximity could represent a nucleation core, however they could be also too far away for a direct interaction. As time passes, the initially denser areas become even more dense and some of the distant molecules induce the formation of a new dense areas (cf. Figure 7.11A). In contrast, some other receptors do not attract the gathering of more molecules, as seen in Figure 7.11B, instead, new small clusters form at another position. Whether such new formed clusters are related to force application cannot be derived from these experiment.

88 CHAPTER 7. SINGLE PARTICLE TRACKING IN LIVING CELLS

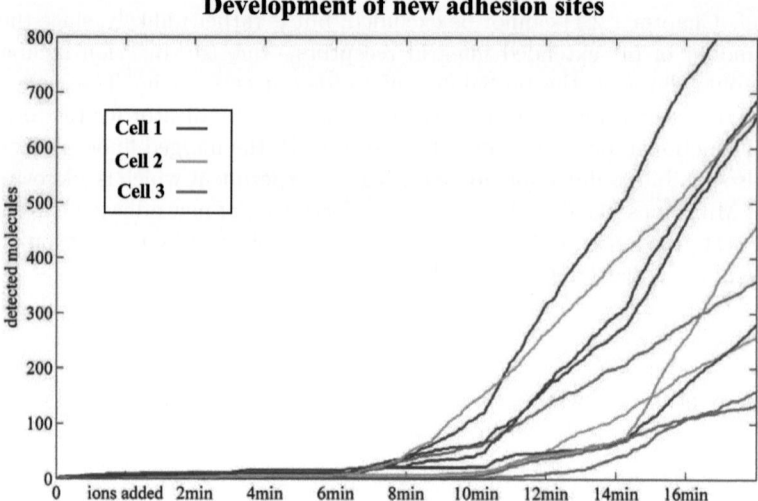

Figure 7.10: Development of new adhesion sites.
Manganese ions trigger the formation of new adhesion sites based on outside-in-activation. The growth curve demonstrates the steady maturation from nascent adhesion into early focal adhesion sites. As the curves do not reach a plateau, the maturation process was not completed when the experiment was stopped.

The final reconstruction of both adhesion sites clearly reveals the aggregated protein structure into dense areas, which was already observed in the PALM-images (cf. Figure 6.7). When interpreting this Figure, the conditions of the measurement must be considered which imply the detection of only a subset of molecules based on labeling, photo-activation and bleaching. Therefore, Figure 7.11 gives only an incomplete picture of the overall situation and is not comparable with the PALM data. Still, as the colorbar was adjusted, regarding this small number of detections, a comparison is justifiable. Also the initial images of nascent adhesion site formation show only a subset of molecules. Therefore, it is possible that also less dense areas in Figure 7.11 could already be part of a cluster which includes receptors that were not activated.

With the chosen two examples in Figure 7.11, it can be further shown that some initially immobilized β3-integrin receptors can be regarded as the seed of a developing dense domain. This suggests diffusion-driven clustering in nascent adhesions as the initial producer of dense domains. However, Figure

7.4. FORMATION OF NEW FOCAL ADHESION SITES

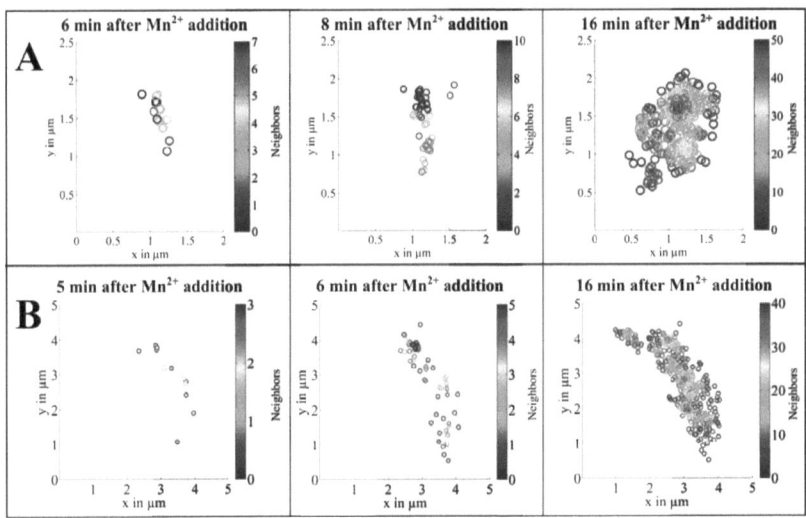

Figure 7.11: Proximity plots of newly formed adhesion sites.
The initial β3-integrin immobilization reveals a broad distribution (B), in which some of the molecules are already nearby, indicated by color (A and B). However, these β3-integrin receptors are mostly too far away for a direct interaction. The final image proves that some of the initially immobile receptors function as the seed of a dense domain (A and B), whereas other dense domains seem to form at a later time point or shift their position (B).

7.11B exhibits that not all initial clusters survive in the process of focal adhesion maturation. The reason cannot be predicted by the performed experiments. A responsibility of traction force in terms of shape and size modulator cannot be ruled out. However, traction force does not function as the initiator, as dense domains appear already in very early time points of adhesion development.

Another remaining question addresses the functionality of protein aggregations. A diffusion-driven formation would rather not support a distinct biological function. Force, as potential shape-modulator, could definitely induce biological signaling processes.

Chapter 8

Dense domains as potential signaling centers

Protein aggregations within focal adhesion sites could function as centers of enhanced signaling activity. Highly active signaling centers should show a distinguishable frequency of protein recruitment compared to less active areas. This could include a higher recruitment rate for proteins having a second messenger function in the signaling machinery. However, also a lower recruitment rate is possible for proteins that need to be bound in order to maintain the signaling events. Single molecule recruitment experiments, using UV spot activation as described in Chapter 5.2.2, can detect higher recruitment rates.

8.1 Spreading efficiency of adhesion proteins

First, a protein with a rather high exchange rate must be found in order to detect enough particles for significant results in the subsequent recruitment experiment. Such experiments are based on precise localization and do not require information about diffusion constants, which enables the use of cytosolic proteins. The cellular spreading efficiency of β3-integrin, Talin, Kindlin2, FAK, Paxillin, Vinculin, Zyxin and Actin was measured, induced by a spatially limited UV-pulse. The spot-activation was performed in a relatively large area of the cell in wide field. Fluorescence events in the whole cell were detected, excluding the activation area. The obtained distribution information is a mix of random blinking and diffusion, as well as exchange and recruitment rate. In Figure 8.1, the results were plotted. As a reference, random blinking without activation was used, and all results normalized accordingly. Random blinking appears in all experiments using photo-activatable fluorophores. A very small fraction of fluorophores is already activated by low-energetic light and produces a fluorescence event. Such a spontaneous activation of PA-fluorophores is a stochastic process and can be considered as random blinking.

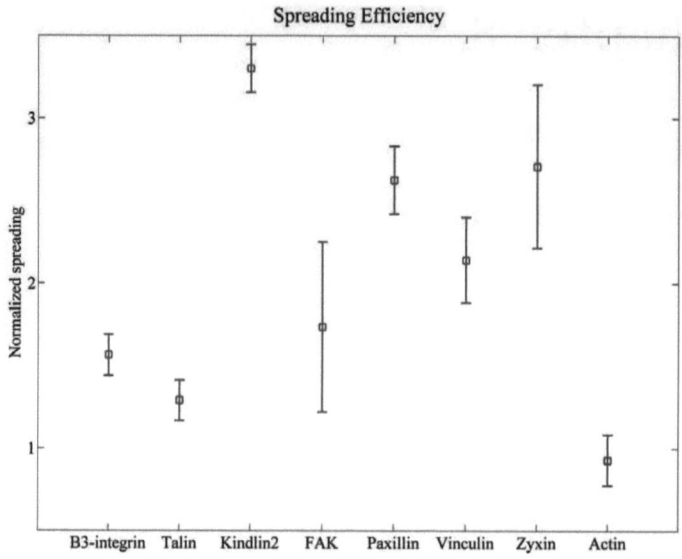

Figure 8.1: Spreading efficiency of adhesion proteins.
Kindlin2 exhibits the highest spreading rate, which suggests a high mobility and exchange rate. Actin, Talin and β3-integrin show only low spreading rates comparable with random blinking. The signaling components Paxillin and FAK, and the scaffold components Vinculin and Zyxin are in between. The large error bars of FAK spreading can be explained by an all-or-nothing spreading efficiency of the individual cells.

Basically no change, compared to random blinking, was observed for Actin. That is not surprising, because big amounts of Actin are assembled into F-actin bundles, which are quite stable [66]. Poor spreading rates were also observed for Talin, as well as for β3-integrin, which is known to have a low exchange rate [11]. Kindlin2 showed a comparably high spreading rate, but spreads with a visible gradient around the activation spot, equally in focal adhesions and in the remaining membrane. This indicates an unspecific recruitment and therefore Kindlin2 was not considered for further experiments. An average distribution rate of 2.5 compared to random blinking was

8.2. RECRUITMENT TO DISTINCT AREAS

observed for the cytoskeletal scaffolding protein Vinculin, which is known for rapid exchange in adhesion sites [37]. Vinculin recruitment correlates with the accessibility of its numerous binding sites inside the rod of Talin, influenced most probably by force. As the results of the sections 7.1 and 7.4 of Chapter 7 suggest a force independent formation of dense domains, the use of Vinculin as a marker for signaling events could induce artifacts. Therefore, Zyxin will also not be further considered, although it shows a proper diffusion rate.

Paxillin is known to be part of the focal adhesion signaling core [76] and belongs to the first binding partners of a nascent adhesion. As its distribution rate is rather high, Paxillin was used for further recruitment experiments. The remaining protein, FAK, also reveals an acceptable distribution rate, however with large error bars. These are explainable by a cell dependent spreading rate, which varies more than usual in cells over-expressing FAK compared to any other adhesion protein. This is an interesting result by itself and could be an indicator for cellular signaling and motility. However, no further experiments were done in this direction. Despite the distribution anomalies, FAK was also used for further experiments. FAK - as a kinase and important part of the signaling complex - could provide important information about recruitment upon signaling in focal adhesion sites.

8.2 FAK and Paxillin recruitment to distinct areas

UV-spot activation was performed to investigate, whether Paxillin and FAK are recruited to distinct domains inside focal adhesions. Random blinking was again used as a reference, before the recruitment to focal adhesions was monitored for two times 1 min. All maturation states of adhesion sites were targeted in a similar way. No obvious gradient was observed between lamellipodia, other cellular edge regions and cell body, which could have been a sign for a directed recruitment. The detected localizations were scanned for neighbors, using degree distribution and a radius of 150 nm. Everything above zero neighbors was defined as clustering and plotted in comparison to random blinking.

The result in Figure 8.2A shows no difference for Paxillin compared to random blinking. As there were no changes observed for 1 or 2 minutes, all results were plotted together. Diagram 8.2B contains the recruitment information of FAK. Here, a pre-screen was performed, considering only cells with a distribution rate higher than 2 compared to random blinking. This procedure excluded already half of the data sets. However, after this cleaning-step, a significant increase of FAK within the set radius of 150 nm was observed.

When comparing both graphs, the rate for random blinking should be considered. For Paxillin this rate is twice as high as for FAK. In contrast, the recruitment rate of both within the set radius is very similar and lies around 50%. This could lead to the assumption that only an artifact was observed, because the recruitment rate of both proteins is practically the same. However, it shoudl be considered, that the random blinking rate makes a statement about the expression level, as a low random blinking corresponds to a low overall expression. In turn, that diminishes also the chance to spot two particles in close proximity, which explains the low reference value in plot 8.2B. As the expression level of FAK was obviously not very high, a proximity recruitment rate of 50% of all molecules 2 min after spot activation should be considered as significantly high.

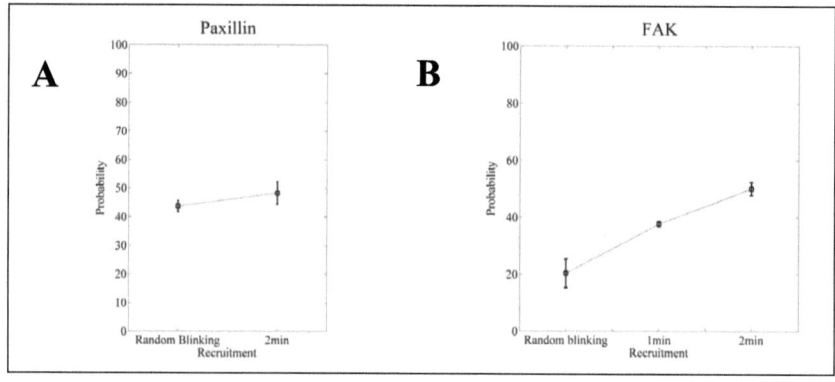

Figure 8.2: Recruitment to dense domains.
Paxillin seems to be equally recruited to all areas of a focal adhesion, as the recruitment rates do not differ from the reference (A). In contrast, FAK has a clear trend towards recruitment into areas of close proximity. The initially activated molecules assemble in a significantly higher amount in a 150 nm radius after 2 min (B).

The questions remains, why Paxillin is not particularly recruited to dense domains, but seems to organize equally in regions of all densities inside adhesion sites. Even though it does not contain a kinetic domain, it is also part of the signaling complex, in which Paxillin becomes highly phosphorylated. Therefore, similar recruitment results for both proteins could be expected. Figure 8.2A already revealed that half of all detected Paxillin molecules in focal adhesions assemble with at least one other Paxillin molecule in a radius

8.2. RECRUITMENT TO DISTINCT AREAS

of 150 nm. An overlay of the wide field image of Paxillin and the recruited particles is visualized in Figure 8.3. Figure 8.3A mirrors the calculated fraction of Figure 8.2A, as there are 11 molecules in close proximity which might correspond to dense domains and 10 molecules which target other areas in the adhesion site. In Figure 8.3B, it seems like Paxillin would be recruited rather to the surrounding area of a focal adhesion, which corresponds to the scaffolding function of Paxillin.

Accordingly, Paxillin could also be recruited to clustered domains, but seems to be an important protein for maintaining the shape of the adhesion site, as well.

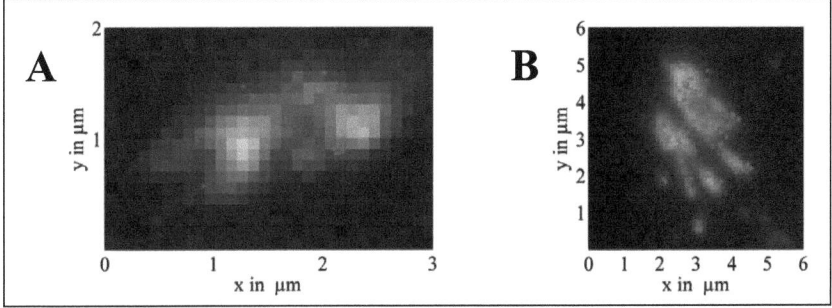

Figure 8.3: Overlay of adhesion site and recruited Paxillin molecules.
Adhesion site A recruits only few Paxillin molecules. Here, a directional recruitment could be assumed, even though the widefield image shows higher intensities in other regions. In Figure B, Paxillin assembles rather in the edge regions of the focal adhesions, which could be an indicator for its function as a scaffold protein.

It should be considered that the observed protein recruitment to clustered domains can simply be a result of probability, caused by density. It is reasonable, that dense domains have a higher fluctuation rate than less dense areas. In these experiments it cannot be distinguished between specific signaling recruitment and usual protein exchange. Therefore, these experiments only show that some adhesion proteins seem to be targeted from distinct domains inside focal adhesions, without answering the question of the purpose. Whether this event is caused by specific functionality or just by probability stays elusive.

Part IV

Discussion and future perspectives

Part 2

Chapter 9

In this work it was shown that adhesion proteins form areas of different densities inside focal adhesion sites. This dynamic process was studied with the adhesion receptor β3-integrin, representing the foundation of a functional adhesion site. The results suggest a force-independent formation of highly dense areas, since the base of a dense area can develop already in the very first steps of adhesion formation. Still, many questions remain unanswered and will be discussed here.

9.1 Nano-organization in fixed cells

In the first part, PALM experiments in fixed cells were performed, exhibiting fundamental information about the intrinsic structure of adhesion sites. The density distribution inside a focal adhesion followed apparently no defined system or polarity and varies a lot in every single adhesion site. Highly dense domains form various shapes and sizes, which complicates the analysis.

9.1.1 Analysis of adhesion protein organization

In this work, a mixture of published software, established algorithms and custom-made mathematical tools was used to study the nano-organization of focal adhesions, which unraveled only a fraction of the actual structure.

The performed characterization of dense domains can give a rough indication of the average size of protein accumulations and their scattering in adhesion sites. However, all clusters were treated equally, although there were large variations in size, shape and density among them. Therefore, no significant differences in terms of cellular localization or protein-dependency were observed. However, it is possible that a more precise definition of cluster-describing parameters could reveal a systematic lateral accumulation. The performed analysis indicates a larger cluster size of some proteins in mature adhesion sites. This was not stated due to the large error bars, originated by merging large and small clusters together. Nevertheless, this preliminary cluster analysis exhibits promising prospects for further steps in terms of cluster organization investigations.

Adhesion protein organization in dense domains

The protein composition inside dense protein accumulations could not be identified, since the degree distributions in Chapter 6.1.2 are based on molecules in a proximity below 50 nm. This applies to all densities, as it was shown that less dense domains also accumulate in this range (cf. Figure 6.8). Therefore, the degree distribution described the average nano-organization of adhesion sites inside the entire adhesion site. However, it can be assumed that the molecular nano-organization varies together with the density.

A reliable analysis of the dense area pattern was not possible with the presented analytical methods. However, the calculated parameters of the degree distribution suggest an individual cluster degree for each protein, supported by highly identical parameters of two randomly distributed controls (Kindlin1 and TMD), which should be further considered. Chapter 6.1.3 proved that the variations in density are mainly independent from the expression level, but are rather based on the particular density of the particular over-expressed protein. Therefore, it can be assumed that a saturated amount of tagged protein localizes in adhesion sites, which enables reliable analysis of the molecular composition. However, it is possible that the efficiency of an over-expressed protein to integrate is reduced by the large fluorescence tag. This is a general problem of fluorescence microscopy and could be counteracted by changing the position or the linker of the fluorophore. Furthermore, there is always a subset of PA-fluorophores that will not be activated, bleached without being detected or re-blinking after being in the triplet state (cf. Chapter 5.1.1). An incomplete detection does automatically influence the parameters of a detected protein cluster and should be considered. However, if there is a distinguishable nanometric organization between protein accumulations and sparse areas inside a focal adhesion, PALM experiments could probably unravel it, even with the problems derived by the fluorescent tags.

However, to do so, the general cluster recognition must be improved, to filter for highly dense domains in order to analyze them separately. Furthermore, the areas of low density should be analyzed, since a lower density could even reveal a higher degree of organization. Obviously, there is room for further improvement of the analysis, which would pave the way for pursuing experiments.

9.1.2 Dual-color PALM

Dense protein aggregations have been observed for each adhesion protein and accordingly hypothesized, that all proteins accumulate together in such dense areas. This assumption seems reasonable, as the initial accumulation of integrin receptors entails also the accumulation of their cytosolic binding partners. Furthermore, most investigated proteins contain a similar amount and size of dense domains in focal adhesions (cf. Table 6.3.2). One method to prove this assumption would be dual-color super-resolution microscopy.

However, the use of several PA-fluorophores in super-resolution microscopy is challenging, as each fluorophore reveals individual activation efficiencies, photon yields and blinking characteristics and requires careful experimental optimization. A localization precision, exhibiting direct protein interactions, is most probably not achievable with the current technologies. However, the simultaneous detection of two different proteins, accumulating in dimensions of 50-100 nm should be feasible.

9.1.3 Protein localization influenced by force inhibition

The effect of traction force inhibition was investigated, and revealed that all cell-matrix proteins remain in proximity to the membrane. However, most probably not only in regions of former adhesion sites but scattered all over the membrane. Similar live cell experiments also exhibit such a remaining membrane localization of Zyxin, even though just very faint (cf. Figure 7.1B). Further cluster analysis of cell-matrix proteins in force inhibited cells suggested a similar nano-cluster behavior as in functional adhesion sites. The only protein exhibiting a clear shrinkage of cluster size upon force inhibition was the receptor β3-integrin. Such observation must be viewed critically, as the fixation treatment is a stressful and time-consuming procedure for cells. The accumulation of membrane-connected proteins in larger structures could represent simply an artifact, caused by membrane alterations due to fixation. However, regardless the questionable clustering, the remaining membrane-connection itself is worth mentioning. Similar accumulations were not observed for adhesion proteins in non-drug treated cells, outside of adhesion sites. Therefore, the membrane-connection seems to be directly related to the relaxation-induced disassembly of adhesion sites.

This could be further studied by inhibiting the general formation of focal adhesions and then investigating the membrane-connection of cell-matrix proteins. It was shown that a poly-lysine coating of the cell dish prevents the formation of adhesion sites [90], due to the lack of suitable ligands. However, the electrostatic interaction of the positive lysine-coating with negatively charged membrane components prevents the cell from detachment [96]. Cells, which are seeded on poly-lysine coatings, are not able to form any adhesion sites and should reveal a similar protein clustering as the non-treated cells, outside of adhesion sites. However, they could also show a similar distribution as the force-inhibited cells, which would entail further discussions about the general localization of inactive cell-matrix proteins in a cell.

9.1.4 Protein localization outside adhesion sites

The scaffold proteins Vinculin and Zyxin have been shown to cluster in a significant amount outside of focal adhesions, even in comparable density fractions (cf. Figure 6.8). The formation of nascent adhesions is unlikely, as the selected areas did not include lamellipodae, and Zyxin does not localize in the nascent adhesion complex. Moreover, the known nascent adhesion components FAK and Talin did not cluster in a comparably large amount. The protein accumulations could represent the remains of a former adhesion site, as it resembles the protein distribution of force inhibited cells. However, the question remains, why such clusters do not appear universally for all cell-matrix proteins.

Since Vinculin and Zyxin show a similar density percentage distribution exclusively outside of adhesion sites, even a direct connection of these proteins is considerable. The LIM-motif of Zyxin is a potential binding site for many proteins [123] and could also be a target for Vinculin. A remaining complex of Zyxin and Vinculin after focal adhesion disassembly could trap these proteins close to the membrane. Zyxin was shown to shuttle between the cytoplasm and the nucleus, as it contains a nuclear export sequence [106]. It is possible, that a nuclear localization of Zyxin leads to a biological impact, which would explain its membrane-maintenance, even after adhesion disassembly. Zyxin and Vinculin could also interact independently from adhesion sites or even build a pre-formed complex, which would describe a new recruitment mechanisms for both proteins.

However, as a direct interaction was not reported so far, *in vitro*-interaction studies should be performed first, before further live cell experiments become conceivable.

9.2 Nano-organization in living cells

The observed protein aggregations have been further studied in live cell experiments with $\beta 3$-integrin as an example. It was shown that $\beta 3$-integrin aggregates do not disassemble upon traction force inhibition, however undergo fluctuations in their position. Therefore, force could function as a modulator rather than an inducer of dense domains. Instead, it was suggested that the seed of a dense domain is already planted in early stages of adhesion development.

9.2.1 Influence of actomyosin contractility

The formation of protein clusters seems to evolve independently from traction force. This assumption is based on the preliminarily results derived from fixed cells (cf. Chapter 6.3.1) and was further confirmed by the results obtained in Chapter 7.2. However, integrin aggregates exhibited alterations in size and shape upon force inhibition. The shrinkage of clustered domains could be explained by the general reduction of molecules in force inhibited focal adhesion sites. However, this cannot explain the changes in the localization of dense domains. A systematic error, caused by the measurement conditions, was suggested. However, Figure 7.6 exhibits the same alterations in lateral positioning of dense domains, even though this experiment was adjusted in order to prevent possible measurement errors. Whether such alterations were directly induced by force inhibition cannot be safely predicted, but such observations did reveal a surprisingly high mobility of dense areas inside focal adhesion sites, as the position varied several hundred nanometers. Preliminary observations in untreated cells reveal similar rearrangements (not shown). However, since the traction force is completely uncontrolled in untreated cells, these observations only support the general high mobility of dense areas without answering the question of force dependency. Since the ECM-exposed ligand is relatively immobile, the density alterations must be caused by the fluctuation rate of the integrin receptor. This means, that the rearrangement of a dense domain must include the formation of new ligand-integrin bindings. Such new connections could be formed by additional, interfusing integrin receptors due to the exchange rate, or by pre-bound integrin receptors, that are released due to force inhibition and rearrange nearby. This assumption seems reasonable, since actomyosin contractility is the main regulator of adhesion sites and could also trigger the rearrangement of integrin receptors.

The Figures 7.5 and 7.6 exhibit rearrangements in a very limited area, which could be explained by the immobility of the ECM ligand. A release of the cytosolic tail-domain of $\beta 3$-integrin from the cytoskeleton does not automatically cause a release from its extracellular ligand, as can be also concluded from Chapter 7.1. Furthermore, Talin exhibited a comparably low disassembly rate upon Y27 incubation, similar to integrin (unpublished observation), which suggests a continuous binding. Therefore, it must be generally assumed that integrin receptors remain in their extended conformation upon force inhibition. The high-affinity, as well as the intermediate state, is directly connected with the immobilization of integrin, as was proven in Figure 7.7C and D. In turn, this means that an eventual extracellular release of integrin could be followed by an immediate re-trapping induced by a suitable ligand in close proximity. This hypothesis leads to a possible rearrangement of dense areas in proximal areas.

In order to test this hypothesis, it would make sense to repeat the same experiments using a dish coating. The ECM component Laminin represents a ligand of some integrin receptors, however it is not the predominant ligand of the $\alpha v \beta 3$-heterodimer. Therefore, the cell could still form functional adhesion sites via the anchoring of its other integrin receptors. $\beta 3$-integrin however, would form less tight bindings which should affect its mobility and in turn the rearrangement of dense areas. The mutant $\beta 3$-integrin$_{primed}$ could be used as well, as it has a mutation in the ligand binding domain. The binding affinity is significantly reduced [117], and can be even more disturbed by using a low affinity Laminin-coating. In both cases, $\beta 3$-integrin release is facilitated, which could affect the rearrangements of dense domains and could even lead to a complete disassembly.

The $\alpha v \beta 3$-integrin receptor has a high binding affinity for Vitronectin, which could be used for dish coating, as well. Here, the position rearrangements of $\beta 3$-integrin should decrease, as an abundant amount of suitable ligand would always be available.

Furthermore, a secondary coating could be considered, whereby a Laminin-coated dish could be spotted with Vitronectin. The Vitronectin dots should be distinguishably spaced without extending a size of 100 nm. Thereby, $\alpha v \beta 3$-integrin could be trapped exclusively in these Vitronectin spots, where they should form a dense domain. If the ligand-receptor-binding is responsible for the maintenance and the re-arrangement of dense domains, those $\alpha v \beta 3$-integrin clusters should only form in a Vitronectin-spotted area and vanish completely upon force-inhibition, instead of re-arrange in neighboring areas.

The investigation of cytosolic adhesion proteins could provide complementary information about force-dependent rearrangements. Their recruitment is based on integrin-tail recognition, which assumes an integrin-dependent clustering. Therefore, the same alterations in positioning of dense domains derived from proximity distribution analysis would be an indication of the integrin-ligand connection as the inducer of the rearrangements. Instead, a rapid disassembly of cytosolic adhesion protein accumulations - leaving an equal distribution - would support force-dependency. A steady disassem-

9.2.2 Diffusion-driven density formation

Figure 7.10 revealed that the formation of a focal adhesion is a steady process, using β3-integrin as an example. No sudden increase of β3-integrin recruitment can be observed within the recorded 8-12 min of the formation process, which could be interpreted as the point of actomyosin connection. This is another argument against force-dependency. Furthermore, Figure 7.11 suggests that the foundation of some dense domains is already determined during nascent adhesion formation. However, it was shown that the initial receptors settle also with quite large distances from each other. The measurement was mentioned to cause problems, as the photo-activated fraction is only small. Distant immobilized receptors could therefore form dense areas as well, but with invisible non-activated or endogenous receptors. Furthermore, there is no continuous control of the current maturation state of the emerging adhesion site, which would contribute to a better interpretation. Figure 7.11B revealed after 6 minutes a highly dense domain in the upper part as well as a growing dense domain in the lower part of the adhesion site. Later, the highly dense domain was still visible, but less dense than some other domains, which were not detected before. The less dense domain was completely disassembled. It is possible that the developing adhesion site was first an accumulation of focal complexes. Some of them survived while others underwent turn-over. However, there is no suitable control for such transition except traction force itself.

β3-integrin is known to be mainly part of mature adhesion sites, while β1-integrin is often localized in the lamellipodia [167]. Generally, both integrin-subunits do accumulate in all maturation states, but it can be expected that nascent adhesions are predominantly formed by heterodimers, containing the β1-subunit. Hence, this protein should be included in the observation of adhesion formation, as its over-expression enhances cell migration in general [167]. The enhanced amount of β1-integrin in nascent adhesions could even reveal a more defined formation of cluster seeds, ending in a dense domain upon adhesion maturation.

The formation of new focal adhesion sites was proven by the recruitment of the focal adhesion marker Zyxin. Only few images of Zyxin were taken during the experiment, in order to not disturb the tracking experiment. However, it would be useful to measure with an alternating dual-color system, which allows the simultaneous detection of the initial receptor clustering indicating a nascent adhesion, and the time point of initial Zyxin recruitment determining the transition into a focal adhesion. This would give the final prove of pre-defined dense domains, as the time point of enhanced actomyosin contractility could be clearly defined.

9.3 Directed protein recruitment

Whether such areas of higher density have a distinct biological function or not, could not be determined with the experiments performed in this work. Figure 8.2B shows FAK recruitment preferentially in distinct areas, compared to the negative control. This could be a sign for enhanced signaling in dense domains, but could also be caused by the exchange rate of FAK. It should be considered, that a recruitment rate of more than 50% FAK molecules in close proximity of a 150 nm radius means in turn a recruitment rate of nearly 50% to other areas, which are not close by. This result is not easy to place, as all given information about dense domains originate from super-resolution experiments. Thereby, the amount of FAK molecules in highly dense domains was below 10%, however the chosen radius was only 25 nm (cf. Figure 6.8).

Generally, most results point to a density formation based on the initial receptor-ligand binding, which should not affect the general ability of signaling. Therefore, a higher amount of signaling events in areas of protein aggregations can be explained simply by a higher amount of proteins available, which potentially induce signaling. Less dense areas contribute to signaling as well, according to their number of proteins. Therefore, the signaling function of dense domains is not considered to be higher than the one of less dense domains. The high amount of FAK recruited to other areas in adhesion sites (cf. Chapter 8.2) indicate a functionality also in areas of low density.

However, a very surprising observation in Chapter 8.2 was not further investigated in this work: the recruitment rate of FAK was highly cell-dependent, meaning that some cells did not show any significant FAK-spreading at all. Such observation could reveal interesting information about the cellular migration state. FAK is a major component of the signaling complex inside a focal adhesion, and was proven to enhance the migration rate of cells by triggering the disassembly of focal adhesions [101]. This should involve a rather high spreading rate of FAK, as disassembly of adhesions necessarily releases FAK and other proteins, while the formation of new adhesion sites requires protein recruitment. In order to address this question, the recruitment rates of FAK as responses to different stimuli could be compared. There are several methods to induce cell migration. Most of them include cell-harming drugs, which create an unfavorable environment and therefore

9.3. DIRECTED PROTEIN RECRUITMENT

promote cellular migration. However, for studies addressing the activation efficiency of FAK, such experiments could yield useful results.

Thereby, a change of cell line should also be considered. In this work, fibroblasts (REF52) and epithelial cells (HeLa) were used, as they are relatively immobile. Furthermore, the spot-activation experiments were performed using Vitronectin-coated dishes. Vitronectin represents the primary ligand of $\alpha v \beta 3$-integrin and supports the strong attachment of mature focal adhesions. Since it was the purpose to detect neighboring molecules in a nanometer proximity, this ligand coating was plausible since it prevented cellular movement and therefore enhances the resolution. However, further experiments should focus more on the dynamic behavior of cells and investigate the directed cell migration, which could even include to observe a possible directed protein recruitment. Such information was not derivable from the experiments performed in Chapter 8.2, as cell migration was prevented. It is certainly conceivable that only particular adhesion sites are targeted by protein recruitment. Pure anchor sites could have lower exchange rates than focal contacts in the cellular edge region. The recruitment efficiency could even reveal information about the direction of cellular migration. Furthermore, an active mutant of FAK [57] could be used, which should enhance cell migration even more.

These results would represent a direct link between cell migration and the dynamic alterations of the signaling process inside focal adhesions and could open new perspectives regarding the sensing function of focal contacts.

Acknowledgments

My years as a PhD student have been a valuable and formative time. I experienced moments of joy alternating with moments of misery. Fortunately, there were always people around to share my feelings with. Therefore, primarily I thank all current and former members of the department 2. I could probably tell a story about every single person, in which I received a helpful advise or shared a good laugh - often even both. Thank you, dear department members for an unforgettable time!

I would like to thank Prof. Philippe Bastiaens for giving me the opportunity to work in his laboratory and thereby to learn many new microscopy techniques.

Prof. Andrea Musacchio assured very spontaneously to be my second examiner - thanks a lot!

I thank Dr. Peter Verveer for funding me and providing me with the crucial equipment to perform my experiments.

In the last months I enjoyed fruitful discussions with Dr. Eli Zamir, who shared his enormous knowledge about focal adhesions with me and gave me further useful suggestions regarding the structure of my thesis. Thanks a lot - I really appreciate this!

Dr. Márton Gelléri, I cannot thank you enough for your support in the last month! Basically, you enabled me to analyze my data the way I wanted - in your free time! Even though we could not realize everything, it was relieving to have someone to count on.

I would like to thank my team members: Jenny Ibach for sharing the PhD time from beginning to the end and being always an imperturbable Zen garden for me. Dr. Yvonne Radon for her useful advices and her willingness to help whenever it is necessary. Thomas Klein for having always a silly verse up his sleeve to put a smile on my face.

I thank Prof. Hernán Grecco for our countless pizza-evenings and even more for providing psychological support when I needed it the most.

The members of my former and my current office should be mentioned here, since they created a friendly working environment and have been always great support. Thank you!

My thanks go further to the members of the cell culture for helping me out in emergency cases of spontaneous microscopy sessions.

Generally, I thank Dr. Astrid Krämer and Tanja Forck, as well as all technicians and engineers for being the organizing scaffold in the background.

Apart from work, I would like to thank my neighbor Helga Werner for providing me with home cooked meals right in front of my door whenever I was coming home late.

References

[1] ABBE, E. Beiträge zur Theorie des Mikroskops und der mikroskopischen Wahrnehmung. *Archiv für Mikroskopische Anatomie 9* (1873), 413–418.

[2] ABERCROMBIE, M., HEAYSMAN, J. E., AND PEGRUM, S. M. The locomotion of fibroblasts in culture. IV. Electron microscopy of the leading lamella. *Experimental cell research 67* (1971), 359–367.

[3] ABRAMOFF, M. D., MAGALHÃES, P. J., RAM, S. J., ABRÀMOFF, M. D., AND HOSPITALS, I. Image processing with ImageJ. *Biophotonics international 11* (2004), 36–42.

[4] ADAIR, B. D., XIONG, J.-P., MADDOCK, C., GOODMAN, S. L., ARNAOUT, M. A., AND YEAGER, M. Three-dimensional EM structure of the ectodomain of integrin {alpha}V{beta}3 in a complex with fibronectin. *The Journal of cell biology 168*, 7 (Mar. 2005), 1109–18.

[5] AIUTI, F., LACAVA, V., GAROFALO, J. A., D'AMELIO, R., AND D'ASERO, C. Surface markers on human lymphocytes: studies of normal subjects and of patients with primary immunodeficiencies. *Clinical and experimental immunology 15* (1973), 43–52.

[6] ALBERTS, B., JOHNSON, A., LEWIS, J., RAFF, M., ROBERTS, K., AND WALTER, P. *Molecular Biology of the Cell, Fourth Edition*, 4th ed. Garland Science, New York, 2002.

[7] ALEKSANDER JABLONSKI. Über den Mechanisms des Photolumineszenz von Farbstoffphosphoren. In *Zeitschrift für Physik*. Springer, 1935, ch. 94, pp. 38–46.

[8] ALEXANDROVA, A. Y., ARNOLD, K., SCHAUB, S., VASILIEV, J. M., MEISTER, J.-J., BERSHADSKY, A. D., AND VERKHOVSKY, A. B. Comparative dynamics of retrograde actin flow and focal adhesions:

formation of nascent adhesions triggers transition from fast to slow flow. *PloS one 3* (2008), e3234.

[9] ANANTHAKRISHNAN, R., AND EHRLICHER, A. The forces behind cell movement. *International journal of biological sciences 3*, 5 (Jan. 2007), 303–17.

[10] ARNOLD, M., CAVALCANTI-ADAM, E. A., GLASS, R., BLÜMMEL, J., ECK, W., KANTLEHNER, M., KESSLER, H., AND SPATZ, J. P. Activation of integrin function by nanopatterned adhesive interfaces. *Chemphyschem : a European journal of chemical physics and physical chemistry 5* (2004), 383–388.

[11] B., W.-H. The Role of Integrins in Cell Migration. In *Madame Curie Bioscience Database*. Landes Bioscience, Austin (TX), 2000.

[12] BACH, C. T. T., SCHEVZOV, G., BRYCE, N. S., GUNNING, P. W., AND O'NEILL, G. M. Tropomyosin isoform modulation of focal adhesion structure and cell migration. *Cell adhesion & migration 4*, 2 (2010), 226–34.

[13] BAKOLITSA, C., COHEN, D. M., BANKSTON, L. A., BOBKOV, A. A., CADWELL, G. W., JENNINGS, L., CRITCHLEY, D. R., CRAIG, S. W., AND LIDDINGTON, R. C. Structural basis for vinculin activation at sites of cell adhesion. *Nature 430*, 6999 (July 2004), 583–6.

[14] BALLESTREM, C., EREZ, N., KIRCHNER, J., KAM, Z., BERSHADSKY, A., AND GEIGER, B. Molecular mapping of tyrosine-phosphorylated proteins in focal adhesions using fluorescence resonance energy transfer. *Journal of Cell Science 119* (2006), 866–875.

[15] BALLESTREM, C., HINZ, B., IMHOF, B. A., AND WEHRLE-HALLER, B. Marching at the front and dragging behind: differential alphaVbeta3-integrin turnover regulates focal adhesion behavior. *The Journal of cell biology 155* (2001), 1319–1332.

[16] BARBAS, C. F. An Allosteric Ca2+ Binding Site on the beta3-Integrins That Regulates the Dissociation Rate for RGD Ligands. *Journal of Biological Chemistry 271*, 36 (Sept. 1996), 21745–21751.

[17] BENINGO, K. A., DEMBO, M., KAVERINA, I., SMALL, J. V., AND WANG, Y. L. Nascent focal adhesions are responsible for the generation of strong propulsive forces in migrating fibroblasts. *The Journal of cell biology 153*, 4 (May 2001), 881–8.

REFERENCES

[18] BETZIG, E., PATTERSON, G. H., SOUGRAT, R., LINDWASSER, O. W., OLENYCH, S., BONIFACINO, J. S., DAVIDSON, M. W., LIPPINCOTT-SCHWARTZ, J., AND HESS, H. F. Imaging intracellular fluorescent proteins at nanometer resolution. *Science (New York, N.Y.) 313*, 5793 (Sept. 2006), 1642–1645.

[19] BOUAOUINA, M., LAD, Y., AND CALDERWOOD, D. A. The N-terminal domains of talin cooperate with the phosphotyrosine binding-like domain to activate beta1 and beta3 integrins. *The Journal of biological chemistry 283* (2008), 6118–6125.

[20] BROUSSARD, J. A., WEBB, D. J., AND KAVERINA, I. Asymmetric focal adhesion disassembly in motile cells. *Current opinion in cell biology 20*, 1 (Mar. 2008), 85–90.

[21] BROWN, M. C., CURTIS, M. S., AND TURNER, C. E. Paxillin LD motifs may define a new family of protein recognition domains. *Nature structural biology 5*, 8 (Aug. 1998), 677–8.

[22] BURRIDGE, K. Foot in mouth: do focal adhesions disassemble by endocytosis? *Nature cell biology 7*, 6 (June 2005), 545–7.

[23] CAI, X., LIETHA, D., CECCARELLI, D. F., KARGINOV, A. V., RAJFUR, Z., JACOBSON, K., HAHN, K. M., ECK, M. J., AND SCHALLER, M. D. Spatial and temporal regulation of focal adhesion kinase activity in living cells. *Molecular and cellular biology 28* (2008), 201–214.

[24] CALDERWOOD, D. A., CAMPBELL, I. D., AND CRITCHLEY, D. R. Talins and kindlins: partners in integrin-mediated adhesion. *Nature reviews. Molecular cell biology 14*, 8 (Aug. 2013), 503–17.

[25] CALDERWOOD, D. A., ULMER, T. S., CRITCHLEY, D., CAMPBELL, I. D., GINSBERG, M. H., LIDDINGTON, R. C., JOLLA, L., AND LE, L. Structural Determinants of Integrin Recognition by Talin. *Ratio 11* (2003), 49–58.

[26] CANCE, W. G., HARRIS, J. E., IACOCCA, M. V., ROCHE, E., YANG, X., CHANG, J., SIMKINS, S., AND XU, L. Immunohistochemical analyses of focal adhesion kinase expression in benign and malignant human breast and colon tissues: correlation with preinvasive and invasive phenotypes. *Clinical cancer research : an official journal of the American Association for Cancer Research 6* (2000), 2417–2423.

[27] CARTER, S. B. Haptotaxis and the mechanism of cell motility. *Nature 213* (1967), 256–260.

[28] CAVALCANTI-ADAM, E. A., VOLBERG, T., MICOULET, A., KESSLER, H., GEIGER, B., AND SPATZ, J. P. Cell spreading and focal adhesion dynamics are regulated by spacing of integrin ligands. *Biophysical Journal 92* (2007), 2964–2974.

[29] CHAN, K. T., BENNIN, D. A., AND HUTTENLOCHER, A. Regulation of adhesion dynamics by calpain-mediated proteolysis of focal adhesion kinase (FAK). *The Journal of biological chemistry 285* (2010), 11418–11426.

[30] CHEN, J., SALAS, A., AND SPRINGER, T. A. Bistable regulation of integrin adhesiveness by a bipolar metal ion cluster. *Nature structural biology 10* (2003), 995–1001.

[31] CHEN, J., AND ZHANG, K. The regulation of integrin function by divalent cations Kun. *Cell Adhesion & Migration 6*, 1 (2012), 20–29.

[32] CHEN, L. M., BAILEY, D., AND FERNANDEZ-VALLE, C. Association of beta 1 integrin with focal adhesion kinase and paxillin in differentiating Schwann cells. *The Journal of neuroscience : the official journal of the Society for Neuroscience 20*, 10 (May 2000), 3776–84.

[33] CHIU, C.-L., AND GRATTON, E. Axial super resolution topography of focal adhesion by confocal microscopy. *Microscopy research and technique 76*, 10 (Oct. 2013), 1070–8.

[34] CHOI, C. K., VICENTE-MANZANARES, M., ZARENO, J., WHITMORE, L. A., MOGILNER, A., AND HORWITZ, A. R. Actin and alpha-actinin orchestrate the assembly and maturation of nascent adhesions in a myosin II motor-independent manner. *Nature cell biology 10* (2008), 1039–1050.

[35] CHOI, C. K., ZARENO, J., DIGMAN, M. A., GRATTON, E., AND HORWITZ, A. R. Cross-correlated fluctuation analysis reveals phosphorylation-regulated paxillin-FAK complexes in nascent adhesions. *Biophysical journal 100* (2011), 583–592.

[36] CLUZEL, C., SALTEL, F. F., LUSSI, J., PAULHE, F. F., IMHOF, B. A., AND WEHRLE-HALLER, B. The mechanisms and dynamics of (alpha)v(beta)3 integrin clustering in living cells. *The Journal of cell biology 171*, 2 (Oct. 2005), 383–92.

[37] COHEN, D. M., KUTSCHER, B., CHEN, H., MURPHY, D. B., AND CRAIG, S. W. A conformational switch in vinculin drives formation and dynamics of a talin-vinculin complex at focal adhesions. *The Journal of biological chemistry 281* (2006), 16006–16015.

[38] CORTESIO, C. L., BOATENG, L. R., PIAZZA, T. M., BENNIN, D. A., AND HUTTENLOCHER, A. Calpain-mediated proteolysis of paxillin negatively regulates focal adhesion dynamics and cell migration. *The Journal of biological chemistry 286* (2011), 9998–10006.

[39] CRITCHLEY, D. R., AND GINGRAS, A. R. Talin at a glance. *Journal of cell science 121*, Pt 9 (May 2008), 1345–7.

[40] DAME, N., ALBERT, R. R., AND BARABÁSI, A.-L. Statistical mechanics of complex networks. *Reviews of modern physics 74*, January (2002), 47–97.

[41] DASZYKOWSKI, M., WALCZAK, B., AND MASSART, D. Looking for natural patterns in data. *Chemometrics and Intelligent Laboratory Systems 56* (2001), 83–92.

[42] DEAKIN, N. O., AND TURNER, C. E. Paxillin comes of age. *Journal of cell science 121*, Pt 15 (Aug. 2008), 2435–44.

[43] DEMALI, K. A., BARLOW, C. A., AND BURRIDGE, K. Recruitment of the Arp2/3 complex to vinculin: coupling membrane protrusion to matrix adhesion. *The Journal of cell biology 159* (2002), 881–891.

[44] DI PAOLO, G., PELLEGRINI, L., LETINIC, K., CESTRA, G., ZONCU, R., VORONOV, S., CHANG, S., GUO, J., WENK, M. R., AND DE CAMILLI, P. Recruitment and regulation of phosphatidylinositol phosphate kinase type 1 gamma by the FERM domain of talin. *Nature 420* (2002), 85–89.

[45] DIGMAN, M. A., BROWN, C. M., HORWITZ, A. R., MANTULIN, W. W., AND GRATTON, E. Paxillin dynamics measured during adhesion assembly and disassembly by correlation spectroscopy. *Biophysical journal 94*, 7 (Apr. 2008), 2819–31.

[46] EFIMOV, A., AND KAVERINA, I. Significance of microtubule catastrophes at focal adhesion sites. *Cell adhesion & migration 3*, 3 (2009), 285–7.

[47] ESTER, M., KRIEGEL, H.-P., XU, X., AND MIINCHEN, D. *A Density-Based Algorithm for Discovering Clusters in Large Spatial Databases with Noise.* AAAI Press, 1996.

[48] ETZIONI, A., DOERSCHUK, C. M., AND HARLAN, J. M. Of man and mouse: leukocyte and endothelial adhesion molecule deficiencies. *Blood 94*, 10 (Nov. 1999), 3281–8.

[49] EZRATTY, E. J., BERTAUX, C., MARCANTONIO, E. E., AND GUNDERSEN, G. G. Clathrin mediates integrin endocytosis for focal adhesion disassembly in migrating cells. *The Journal of cell biology 187*, 5 (Nov. 2009), 733–47.

[50] EZRATTY, E. J., PARTRIDGE, M. A., AND GUNDERSEN, G. G. Microtubule-induced focal adhesion disassembly is mediated by dynamin and focal adhesion kinase. *Nature cell biology 7*, 6 (June 2005), 581–90.

[51] FOURNIER, M. F., SAUSER, R., AMBROSI, D., MEISTER, J.-J., AND VERKHOVSKY, A. B. Force transmission in migrating cells. *The Journal of cell biology 188* (2010), 287–297.

[52] FRAME, M. C., PATEL, H., SERRELS, B., LIETHA, D., AND ECK, M. J. The FERM domain: organizing the structure and function of FAK. *Nature Reviews Molecular Cell Biology 11*, 11 (Nov. 2010), 802–14.

[53] FRANCO, S. J., RODGERS, M. A., PERRIN, B. J., HAN, J., BENNIN, D. A., CRITCHLEY, D. R., AND HUTTENLOCHER, A. Calpain-mediated proteolysis of talin regulates adhesion dynamics. *Nature cell biology 6* (2004), 977–983.

[54] FRANTZ, C., STEWART, K. M., AND WEAVER, V. M. The extracellular matrix at a glance. *Journal of cell science 123*, Pt 24 (Dec. 2010), 4195–200.

[55] FRANZ, C. M., AND MÜLLER, D. J. Analyzing focal adhesion structure by atomic force microscopy. *Journal of cell science 118* (2005), 5315–5323.

[56] FREED, E., GAILIT, J., AND GEER, P. V. D. novel integrin beta subunit is associated with the vitronectin receptor alpha subunit (alpha v) in a human osteosarcoma cell line and is a substrate for protein kinase C. *The EMBO Journal 8*, 1 (1989), 2955–2965.

[57] GABARRA-NIECKO, V., KEELY, P. J., AND SCHALLER, M. D. Characterization of an activated mutant of focal adhesion kinase: 'Super-FAK'. *The Biochemical journal 365*, Pt 3 (Aug. 2002), 591–603.

[58] GAILLARD, T., DEJAEGERE, A., AND STOTE, R. H. Dynamics of beta3 integrin I-like and hybrid domains: insight from simulations on the mechanism of transition between open and closed forms. *Proteins 76* (2009), 977–994.

[59] GALBRAITH, C. G., YAMADA, K. M., AND GALBRAITH, J. A. Polymerizing actin fibers position integrins primed to probe for adhesion sites. *Science (New York, N.Y.) 315*, 5814 (Feb. 2007), 992–5.

[60] GARDEL, M. L., SCHNEIDER, I. C., ARATYN-SCHAUS, Y., AND WATERMAN, C. M. Mechanical integration of actin and adhesion dynamics in cell migration. *Annual review of cell and developmental biology 26* (Jan. 2010), 315–33.

[61] GEIGER, B., SPATZ, J. P., AND BERSHADSKY, A. D. Environmental sensing through focal adhesions. *Nature reviews. Molecular cell biology 10*, 1 (Jan. 2009), 21–33.

[62] GLANZMANN, E. Hereditäre hämorrhägische Thrombasthenie. Ein Beitrag zur Pathologie der Blutplättchen. In *Jahrbuch für Kinderheilkunde 88*. 1918, pp. 1–42, 113–141.

[63] GRINNELL, F., AND BACKMAN, R. Role of Integrin Receptors in Manganese-Dependent Cell Spreading on Albumin-Coated Substrata. *Experimental Cell Research 223* (1991), 218–223.

[64] GURSKAYA, N. G., VERKHUSHA, V. V., SHCHEGLOV, A. S., STAROVEROV, D. B., CHEPURNYKH, T. V., FRADKOV, A. F., LUKYANOV, S., AND LUKYANOV, K. A. Engineering of a monomeric green-to-red photoactivatable fluorescent protein induced by blue light. *Nature biotechnology 24*, 4 (Apr. 2006), 461–5.

[65] HANNA, S., AND EL-SIBAI, M. Signaling networks of Rho GTPases in cell motility. *Cellular signalling 25* (2013), 1955–61.

[66] HARDY, L. R. Fluorescence recovery after photobleaching (FRAP) with a focus on F-actin. In *Current protocols in neuroscience*, J. N. Crawley, Ed., vol. Chapter 2. Wiley Online Library, Jan. 2012, ch. 2, p. Unit 2.17.

[67] HESS, S. T., GIRIRAJAN, T. P. K., AND MASON, M. D. Ultra-high resolution imaging by fluorescence photoactivation localization microscopy. *Biophysical Journal 91*, 11 (2006), 4258–4272.

[68] HU, P., AND LUO, B.-H. Integrin bi-directional signaling across the plasma membrane. *Journal of cellular physiology 228*, 2 (Feb. 2013), 306–12.

[69] HYNES, R. O. Cell adhesion: old and new questions. *Trends in cell biology 9*, 12 (Dec. 1999), M33–7.

[70] IZAGUIRRE, G., AGUIRRE, L., HU, Y. P., LEE, H. Y., SCHLAEPFER, D. D., ANESKIEVICH, B. J., AND HAIMOVICH, B. The cytoskeletal/non-muscle isoform of alpha-actinin is phosphorylated on its actin-binding domain by the focal adhesion kinase. *The Journal of biological chemistry 276*, 31 (Aug. 2001), 28676–85.

[71] IZZARD, C. S., AND LOCHNER, L. R. Cell-to-substrate contacts in living fibroblasts: an interference reflexion study with an evaluation of the technique. *Journal of cell science 21*, 1 (June 1976), 129–59.

[72] JAMIESON, J. S., TUMBARELLO, D. A., HALLÉ, M., BROWN, M. C., TREMBLAY, M. L., AND TURNER, C. E. Paxillin is essential for PTP-PEST-dependent regulation of cell spreading and motility: a role for paxillin kinase linker. *Journal of cell science 118* (2005), 5835–5847.

[73] JAQAMAN, K., LOERKE, D., METTLEN, M., KUWATA, H., GRINSTEIN, S., SCHMID, S. L., AND DANUSER, G. Robust single-particle tracking in live-cell time-lapse sequences. *Nature methods 5* (2008), 695–702.

[74] KAHNER, B. N., KATO, H., BANNO, A., GINSBERG, M. H., SHATTIL, S. J., AND YE, F. Kindlins, integrin activation and the regulation of talin recruitment to $\alpha IIb\beta 3$. *PloS one 7*, 3 (Jan. 2012), e34056.

[75] KALLI, A. C., WEGENER, K. L., GOULT, B. T., ANTHIS, N. J., CAMPBELL, I. D., AND SANSOM, M. S. P. The structure of the talin/integrin complex at a lipid bilayer: an NMR and MD simulation study. *Structure (London, England : 1993) 18*, 10 (Oct. 2010), 1280–8.

[76] KANCHANAWONG, P., SHTENGEL, G., PASAPERA, A. M., RAMKO, E. B., DAVIDSON, M. W., HESS, H. F., AND WATERMAN, C. M.

Nanoscale architecture of integrin-based cell adhesions. *Nature 468*, 7323 (Nov. 2010), 580–584.

[77] KATZ, B. Z., ZAMIR, E., BERSHADSKY, A., KAM, Z., YAMADA, K. M., AND GEIGER, B. Physical state of the extracellular matrix regulates the structure and molecular composition of cell-matrix adhesions. *Molecular biology of the cell 11* (2000), 1047–1060.

[78] KAVERINA, I., KRYLYSHKINA, O., AND SMALL, J. V. Microtubule Targeting of Substrate Contacts Promotes Their Relaxation. *The Journal of Cell Biology 146*, 5 (1999), 1033–1043.

[79] KIOSSES, W. B., SHATTIL, S. J., PAMPORI, N., AND SCHWARTZ, M. A. Rac recruits high-affinity integrin alphavbeta3 to lamellipodia in endothelial cell migration. *Nature cell biology 3* (2001), 316–320.

[80] KONG, F., GARCÍA, A. J., MOULD, A. P., HUMPHRIES, M. J., AND ZHU, C. Demonstration of catch bonds between an integrin and its ligand. *The Journal of cell biology 185*, 7 (June 2009), 1275–84.

[81] LAUDANNA, C., KIM, J. Y., CONSTANTIN, G., AND BUTCHER, E. Rapid leukocyte integrin activation by chemokines. *Immunological reviews 186* (Aug. 2002), 37–46.

[82] LAWSON, C., LIM, S.-T. S.-T., URYU, S., CHEN, X. L., CALDERWOOD, D. A., AND SCHLAEPFER, D. D. FAK promotes recruitment of talin to nascent adhesions to control cell motility. *The Journal of cell biology 196*, 2 (Jan. 2012), 223–32.

[83] LAWSON, C., AND SCHLAEPFER, D. D. Integrin adhesions: who's on first? What's on second? Connections between FAK and talin. *Cell adhesion & migration 6*, 4 (2012), 302–6.

[84] LEE, S. Y., VORONOV, S., LETINIC, K., NAIRN, A. C., DI PAOLO, G., AND DE CAMILLI, P. Regulation of the interaction between PIPKI gamma and talin by proline-directed protein kinases. *The Journal of cell biology 168*, 5 (Feb. 2005), 789–99.

[85] LEGATE, K. R., AND FÄSSLER, R. Mechanisms that regulate adaptor binding to beta-integrin cytoplasmic tails. *Journal of cell science 122*, Pt 2 (Jan. 2009), 187–98.

[86] LEGATE, K. R., WICKSTRÖM, S. A., AND FÄSSLER, R. Genetic and cell biological analysis of integrin outside-in signaling. *Genes & development 23*, 4 (Mar. 2009), 397–418.

[87] LIDDINGTON, R. C., AND GINSBERG, M. H. Integrin activation takes shape. *The Journal of cell biology 158*, 5 (Sept. 2002), 833–9.

[88] LITVINOV, R. I., SHUMAN, H., BENNETT, J. S., AND WEISEL, J. W. Binding strength and activation state of single fibrinogen-integrin pairs on living cells. *Proceedings of the National Academy of Sciences of the United States of America 99*, 11 (May 2002), 7426–31.

[89] LIU, X., AND SCHNELLMANN, R. Calpain mediates progressive plasma membrane permeability and proteolysis of cytoskeleton-associated paxillin, talin, and vinculin during renal cell death. *Journal of Pharmacology and Experimental Therapeutics 304*, 1 (2003), 63–70.

[90] LO, Y. Y. C. Requirements of Focal Adhesions and Calcium Fluxes for Interleukin-1-induced ERK Kinase Activation and c-fos Expression in Fibroblasts. *Journal of Biological Chemistry 273*, 12 (Mar. 1998), 7059–7065.

[91] LUO, B.-H., CARMAN, C. V., AND SPRINGER, T. A. Structural basis of integrin regulation and signaling. *Annual review of immunology 25* (2007), 619–647.

[92] MA, Y.-Q., QIN, J., WU, C., AND PLOW, E. F. Kindlin-2 (Mig-2): a co-activator of beta3 integrins. *The Journal of cell biology 181*, 3 (May 2008), 439–46.

[93] MAROUDAS, N. G. Adhesion and Spreading of Cells on Charged Surfaces. *Journal of theoretical Biology, 123* (1975), 417–424.

[94] MASTERS, B. R., AND LAKOWICZ, J. R. Principles of Fluorescence Spectroscopy, Third Edition, 2008.

[95] MATTHEYSES, A. L., SIMON, S. M., AND RAPPOPORT, J. Z. Imaging with total internal reflection fluorescence microscopy for the cell biologist. *Journal of Cell Science 123*, Pt 21 (Nov. 2010), 3621–3628.

[96] MAZIA, D., SCHATTEN, G., AND WINFIELD, S. Adhesion of cells to surfaces coated with polylysine. *Journal of cell biology 66*, 3 (1975), 198–200.

[97] MCKINNEY, S. A., MURPHY, C. S., HAZELWOOD, K. L., DAVIDSON, M. W., AND LOOGER, L. L. A bright and photostable photoconvertible fluorescent protein. *Nature methods 6*, 2 (2009), 131–133.

REFERENCES

[98] MCLEAN, G. W., CARRAGHER, N. O., AVIZIENYTE, E., EVANS, J., BRUNTON, V. G., AND FRAME, M. C. The role of focal-adhesion kinase in cancer - a new therapeutic opportunity. *Nature reviews. Cancer 5*, 7 (July 2005), 505–15.

[99] MIAO, H., LI, S., HU, Y.-L., YUAN, S., ZHAO, Y., CHEN, B. P. C., PUZON-MCLAUGHLIN, W., TARUI, T., SHYY, J. Y.-J., TAKADA, Y., USAMI, S., AND CHIEN, S. Differential regulation of Rho GTPases by beta1 and beta3 integrins: the role of an extracellular domain of integrin in intracellular signaling. *Journal of cell science 115*, Pt 10 (May 2002), 2199–206.

[100] MICHALET, X. Mean square displacement analysis of single-particle trajectories with localization error: Brownian motion in an isotropic medium. *Physical review. E, Statistical, nonlinear, and soft matter physics 82* (2010), 041914.

[101] MITRA, S. K., HANSON, D. A., AND SCHLAEPFER, D. D. Focal adhesion kinase: in command and control of cell motility. *Nature reviews. Molecular cell biology 6*, 1 (Jan. 2005), 56–68.

[102] MIYAMOTO, S., AKIYAMA, S. K., AND YAMADA, K. M. Synergistic roles for receptor occupancy and aggregation in integrin transmembrane function. *Science (New York, N.Y.) 267*, 5199 (Feb. 1995), 883–5.

[103] MIZUNO, H., MAL, T. K., TONG, K. I., ANDO, R., FURUTA, T., IKURA, M., AND MIYAWAKI, A. Photo-induced peptide cleavage in the green-to-red conversion of a fluorescent protein. *Molecular cell 12* (2003), 1051–1058.

[104] NAYAL, A., WEBB, D. J., AND HORWITZ, A. F. Talin: an emerging focal point of adhesion dynamics. *Current opinion in cell biology 16*, 1 (Feb. 2004), 94–8.

[105] NIU, G., AND CHEN, X. Why integrin as a primary target for imaging and therapy. *Theranostics 1* (Jan. 2011), 30–47.

[106] NIX, D. A., AND BECKERLE, M. C. Nuclear-cytoplasmic shuttling of the focal contact protein, zyxin: a potential mechanism for communication between sites of cell adhesion and the nucleus. *The Journal of cell biology 138* (1997), 1139–1147.

[107] OSER, M., AND CONDEELIS, J. The cofilin activity cycle in lamellipodia and invadopodia. *Journal of cellular biochemistry 108* (2009), 1252–1262.

[108] PAPUSHEVA, E., MELLO DE QUEIROZ, F., DALOUS, J., HAN, Y., ESPOSITO, A., JARES-ERIJMANXA, E. A., JOVIN, T. M., AND BUNT, G. Dynamic conformational changes in the FERM domain of FAK are involved in focal-adhesion behavior during cell spreading and motility. *Journal of cell science 122*, Pt 5 (Mar. 2009), 656–66.

[109] PATTERSON, G. H., AND LIPPINCOTT-SCHWARTZ, J. A photoactivatable GFP for selective photolabeling of proteins and cells. *Science (New York, N.Y.) 297*, 5588 (Sept. 2002), 1873–7.

[110] PIATKEVICH, K. D., AND VERKHUSHA, V. V. Advances in engineering of fluorescent proteins and photoactivatable proteins with red emission. *Current opinion in chemical biology 14* (2010), 23–29.

[111] POLLARD, T. D., AND BORISY, G. G. Cellular motility driven by assembly and disassembly of actin filaments. *Cell 112*, 4 (Feb. 2003), 453–65.

[112] PUKLIN-FAUCHER, E., AND SHEETZ, M. P. The mechanical integrin cycle. *Journal of Cell Science 122*, 4 (Feb. 2009), 575–575.

[113] RABORN, J., WANG, W., AND LUO, B.-H. Regulation of integrin $\alpha IIb\beta 3$ ligand binding and signaling by the metal ion binding sites in the β I domain. *Biochemistry 50* (2011), 2084–2091.

[114] RAYMO, F. M. Photoactivatable Fluorophores. *ISRN Physical Chemistry 2012* (2012), 1–15.

[115] ROCCO, M., ROSANO, C., WEISEL, J. W., HORITA, D. A., AND HANTGAN, R. R. Integrin conformational regulation: uncoupling extension/tail separation from changes in the head region by a multiresolution approach. *Structure (London, England : 1993) 16*, 6 (June 2008), 954–64.

[116] ROLAND WINTER & FRANK NOLL. *Methoden der Biophysikalischen Chemie*, 1 ed. Teubner, Stuttgart, 1998.

[117] ROSSIER, O., OCTEAU, V., AND SIBARITA, J. Integrins $\beta 1$ and $\beta 3$ exhibit distinct dynamic nanoscale organizations inside focal adhesions. *Nature cell biology 14*, 10 (Oct. 2012), 1057–67.

[118] RUDOLPH, R., AND CHERESH, D. Cell adhesion mechanisms and their potential impact on wound healing and tumor control. *Clinics in plastic surgery 17* (1990), 457–462.

[119] RUST, M. J., BATES, M., AND ZHUANG, X. Sub-diffraction-limit imaging by stochastic optical reconstruction microscopy (STORM). *Nature Methods 3*, 10 (2006), 793–795.

[120] SAINIO, A., KOULU, M., WIGHT, T. N., PENTTINEN, R., AND JA, H. Extracellular Matrix Molecules : Potential Targets in Pharmacotherapy. *Pharmacological Review 61*, 2 (2009), 198–223.

[121] SCHALLER, M. D., OTEY, C. A., HILDEBRAND, J. D., AND PARSONS, J. T. Focal adhesion kinase and paxillin bind to peptides mimicking beta integrin cytoplasmic domains. *The Journal of cell biology 130* (1995), 1181–1187.

[122] SCHLAEPFER, D. D., JONES, K. C., AND HUNTER, T. Multiple Grb2-mediated integrin-stimulated signaling pathways to ERK2/mitogen-activated protein kinase: summation of both c-Src- and focal adhesion kinase-initiated tyrosine phosphorylation events. *Molecular and cellular biology 18* (1998), 2571–2585.

[123] SCHMEICHEL, K. L., AND BECKERLE, M. C. The LIM domain is a modular protein-binding interface. *Cell 79* (1994), 211–219.

[124] SCHRAMP, M., AND HEDMAN, A. *Phosphoinositides I: Enzymes of Synthesis and Degradation*. Springer Netherlands, Dodrecht, 2012.

[125] SELIGSOHN, U. Glanzmann thrombasthenia: a model disease which paved the way to powerful therapeutic agents. *Pathophysiology of haemostasis and thrombosis 32*, 5-6 (2002), 216–7.

[126] SERRELS, B., SERRELS, A., BRUNTON, V. G., HOLT, M., MCLEAN, G. W., GRAY, C. H., JONES, G. E., AND FRAME, M. C. Focal adhesion kinase controls actin assembly via a FERM-mediated interaction with the Arp2/3 complex. *Nature cell biology 9* (2007), 1046–1056.

[127] SHAN, Y., YU, L., LI, Y., PAN, Y., ZHANG, Q., WANG, F., CHEN, J., AND ZHU, X. Nudel and FAK as antagonizing strength modulators of nascent adhesions through paxillin. *PLoS biology 7* (2009), e1000116.

[128] SHATTIL, S. J., KIM, C., AND GINSBERG, M. H. The final steps of integrin activation: the end game. *Nature Reviews Molecular Cell Biology 11* (2010), 288–300.

[129] SHERRATT, J. A., AND MURRAY, J. D. Models of epidermal wound healing. *Proceedings of The Royal Society 241* (1990), 29–36.

[130] SHIMAOKA, M., TAKAGI, J., AND SPRINGER, T. A. Conformational regulation of integrin structure and function. *Annual review of biophysics and biomolecular structure 31* (Jan. 2002), 485–516.

[131] SHROFF, H., GALBRAITH, C. G., GALBRAITH, J. A., AND BETZIG, E. Live-cell photoactivated localization microscopy of nanoscale adhesion dynamics. *Nature 5*, 5 (2008), 417–423.

[132] SHROFF, H., GALBRAITH, C. G., GALBRAITH, J. A., WHITE, H., GILLETTE, J., OLENYCH, S., DAVIDSON, M. W., AND BETZIG, E. Dual-color superresolution imaging of genetically expressed probes within individual adhesion complexes. *Proceedings of the National Academy of Sciences of the United States of America 104*, 51 (2007), 20308–20313.

[133] SHROFF, H., WHITE, H., AND BETZIG, E. Photoactivated localization microscopy (PALM) of adhesion complexes. In *Current protocols in cell biology*, J. S. Bonifacino, Ed., vol. Chapter 4. Wiley Online Library, Dec. 2008, ch. 4, p. Unit 4.21.

[134] SMITH, J. W., PIOTROWICZ, R. S., AND MATHIS, D. A mechanism for divalent cation regulation of beta 3-integrins. *The Journal of biological chemistry 269* (1994), 960–967.

[135] SONG, X., YANG, J., HIRBAWI, J., YE, S., PERERA, H. D., GOKSOY, E., DWIVEDI, P., PLOW, E. F., ZHANG, R., AND QIN, J. A novel membrane-dependent on/off switch mechanism of talin FERM domain at sites of cell adhesion. *Cell research 22*, 11 (Nov. 2012), 1533–45.

[136] STEPANENKO, O. V., STEPANENKO, O. V., SHCHERBAKOVA, D. M., KUZNETSOVA, I. M., TUROVEROV, K. K., AND VERKHUSHA, V. V. Modern fluorescent proteins: from chromophore formation to novel intracellular applications. *BioTechniques 51*, 5 (Nov. 2011), 313–4, 316, 318.

REFERENCES

[137] TAKAGI, J., PETRE, B. M., WALZ, T., AND SPRINGER, T. A. Global conformational rearrangements in integrin extracellular domains in outside-in and inside-out signaling. *Cell 110*, 5 (Sept. 2002), 599–11.

[138] TAMKUN, J. W., DESIMONE, D. W., FONDA, D., PATEL, R. S., BUCK, C., HORWITZ, A. F., AND HYNES, R. O. Structure of integrin, a glycoprotein involved in the transmembrane linkage between fibronectin and actin. *Cell 46* (1986), 271–282.

[139] TANAKA, K. A. K., SUZUKI, K. G. N., SHIRAI, Y. M., SHIBUTANI, S. T., MIYAHARA, M. S. H., TSUBOI, H., YAHARA, M., YOSHIMURA, A., MAYOR, S., FUJIWARA, T. K., AND KUSUMI, A. Membrane molecules mobile even after chemical fixation. *Nature methods 7*, 11 (Nov. 2010), 865–6.

[140] TEN KLOOSTER, J. P., AND HORDIJK, P. L. Targeting and localized signalling by small GTPases. *Biology of the cell 99*, 1 (Jan. 2007), 1–12.

[141] THERIOT, J. A., AND MITCHISON, T. J. Comparison of actin and cell surface dynamics in motile fibroblasts. *The Journal of cell biology 119* (1992), 367–377.

[142] TURNER, C. E. Paxillin interactions. *Journal of cell science 113 Pt 23* (2000), 4139–4140.

[143] UEHATA, M., ISHIZAKI, T., SATOH, H., ONO, T., KAWAHARA, T., MORISHITA, T., TAMAKAWA, H., YAMAGAMI, K., INUI, J., MAEKAWA, M., AND NARUMIYA, S. Calcium sensitization of smooth muscle mediated by a Rho-associated protein kinase in hypertension. *Nature 389*, 6654 (Oct. 1997), 990–4.

[144] VICENTE-MANZANARES, M., CHOI, C. K., AND HORWITZ, A. R. Integrins in cell migration–the actin connection. *Journal of cell science 122* (2009), 199–206.

[145] VICENTE-MANZANARES, M., AND HORWITZ, A. R. Adhesion dynamics at a glance. *Journal of cell science 124*, Pt 23 (Dec. 2011), 3923–7.

[146] VINOGRADOVA, O., VELYVIS, A., VELYVIENE, A., HU, B., HAAS, T. A., PLOW, E. F., AND QIN, J. A Structural Mechanism of Integrin alpha IIb beta3 Inside-Out Activation as Regulated by Its Cytoplasmic Face. *Cell 110* (2002), 587–597.

[147] WAKATSUKI, T., SCHWAB, B., THOMPSON, N. C., AND ELSON, E. L. Effects of cytochalasin D and latrunculin B on mechanical properties of cells. *Journal of cell science 114*, Pt 5 (Mar. 2001), 1025–36.

[148] WANG, J.-H. Pull and push: talin activation for integrin signaling. *Cell research 22*, 11 (Nov. 2012), 1512–4.

[149] WANG, Y., CAO, H., CHEN, J., AND MCNIVEN, M. A. A direct interaction between the large GTPase dynamin-2 and FAK regulates focal adhesion dynamics in response to active Src. *Molecular biology of the cell 22* (2011), 1529–1538.

[150] WEBB, D. J., DONAIS, K., WHITMORE, L. A., THOMAS, S. M., TURNER, C. E., PARSONS, J. T., AND HORWITZ, A. F. FAK-Src signalling through paxillin, ERK and MLCK regulates adhesion disassembly. *Nature Cell Biology 6*, 2 (2004), 154–161.

[151] WEGENER, K. L., PARTRIDGE, A. W., HAN, J., PICKFORD, A. R., LIDDINGTON, R. C., GINSBERG, M. H., AND CAMPBELL, I. D. Structural basis of integrin activation by talin. *Cell 128*, 1 (Jan. 2007), 171–82.

[152] WEHRLE-HALLER, B. Assembly and disassembly of cell matrix adhesions. *Current opinion in cell biology 24*, 5 (Oct. 2012), 569–81.

[153] WEHRLE-HALLER, B. Structure and function of focal adhesions. *Current opinion in cell biology 24*, 1 (Mar. 2012), 116–24.

[154] WELF, E. S., NAIK, U. P., AND OGUNNAIKE, B. A. A spatial model for integrin clustering as a result of feedback between integrin activation and integrin binding. *Biophysical journal 103*, 6 (Sept. 2012), 1379–89.

[155] WIEDENMANN, J., IVANCHENKO, S., OSWALD, F., SCHMITT, F., RÖCKER, C., SALIH, A., SPINDLER, K.-D., AND NIENHAUS, G. U. EosFP, a fluorescent marker protein with UV-inducible green-to-red fluorescence conversion. *Proceedings of the National Academy of Sciences of the United States of America 101*, 45 (Nov. 2004), 15905–10.

[156] WOLF, E. P. Experimental studies on inflammation: The infuence of chemicals upon the chemotaxis of leucocytes in vitro. *The Journal of experimental medicine 34* (1921), 375–396.

[157] WOLFENSON, H., HENIS, Y. I., GEIGER, B., AND BERSHADSKY, A. D. The heel and toe of the cell's foot: a multifaceted approach

for understanding the structure and dynamics of focal adhesions. *Cell motility and the cytoskeleton 66*, 11 (Nov. 2009), 1017–29.

[158] WOLTER, S., LÖSCHBERGER, A., HOLM, T., AUFMKOLK, S., DABAUVALLE, M.-C., VAN DE LINDE, S., AND SAUER, M. rapid-STORM: accurate, fast open-source software for localization microscopy. *Nature methods 9* (2012), 1040–41.

[159] WOZNIAK, M. A., MODZELEWSKA, K., KWONG, L., AND KEELY, P. J. Focal adhesion regulation of cell behavior. *Biochimica et biophysica acta 1692*, 2-3 (July 2004), 103–19.

[160] WU, X., SUETSUGU, S., COOPER, L. A., TAKENAWA, T., AND GUAN, J. L. Focal adhesion kinase regulation of N-WASP subcellular localization and function. *Journal of Biological Chemistry 279* (2004), 9565–9576.

[161] XIAO, T., TAKAGI, J., COLLER, B. S., WANG, J.-H., AND SPRINGER, T. A. Structural basis for allostery in integrins and binding to fibrinogen-mimetic therapeutics. *Nature 432*, 7013 (Nov. 2004), 59–67.

[162] XIONG, J., BALCIOGLU, H. E., AND DANEN, E. H. J. Integrin signaling in control of tumor growth and progression. *The international journal of biochemistry & cell biology 45* (2013), 1012–15.

[163] XIONG, J.-P., MAHALINGHAM, B., ALONSO, J. L., BORRELLI, L. A., RUI, X., ANAND, S., HYMAN, B. T., RYSIOK, T., MÜLLER-POMPALLA, D., GOODMAN, S. L., AND ARNAOUT, M. A. Crystal structure of the complete integrin $\alpha V\beta 3$ ectodomain plus an α/β transmembrane fragment. *The Journal of Cell Biology 186* (2009), 589–600.

[164] XIONG, J. P., STEHLE, T., DIEFENBACH, B., ZHANG, R., DUNKER, R., SCOTT, D. L., JOACHIMIAK, A., GOODMAN, S. L., AND ARNAOUT, M. A. Crystal structure of the extracellular segment of integrin alpha Vbeta3. *Science (New York, N.Y.) 294*, 5541 (Oct. 2001), 339–45.

[165] XIONG, J.-P., STEHLE, T., ZHANG, R., JOACHIMIAK, A., FRECH, M., GOODMAN, S. L., AND ARNAOUT, M. A. Crystal structure of the extracellular segment of integrin alpha Vbeta3 in complex with an Arg-Gly-Asp ligand. *Science (New York, N.Y.) 296* (2002), 151–155.

[166] YATES, L. A., LUMB, C. N., BRAHME, N. N., ZALYTE, R., BIRD, L. E., DE COLIBUS, L., OWENS, R. J., CALDERWOOD, D. A., SANSOM, M. S. P., AND GILBERT, R. J. C. Structural and functional characterization of the kindlin-1 pleckstrin homology domain. *The Journal of biological chemistry 287* (2012), 43246–61.

[167] ZAIDEL-BAR, R. Job-splitting among integrins. *Nature cell biology 15*, 6 (June 2013), 575–7.

[168] ZAMIR, E., AND GEIGER, B. Molecular complexity and dynamics of cell-matrix adhesions. *Journal of cell science 114*, Pt 20 (Oct. 2001), 3583–3590.

[169] ZAMIR, E., KATZ, M., POSEN, Y., EREZ, N., YAMADA, K. M., KATZ, B. Z., LIN, S., LIN, D. C., BERSHADSKY, A., KAM, Z., AND GEIGER, B. Dynamics and segregation of cell-matrix adhesions in cultured fibroblasts. *Nature cell biology 2* (2000), 191–196.

[170] ZENT, R., AND POZZI, A., Eds. *Cell-Extracellular Matrix Interactions in Cancer*. Springer New York, New York, NY, 2010.

[171] ZHENG, Y., XIA, Y., HAWKE, D., HALLE, M., TREMBLAY, M. L., GAO, X., ZHOU, X. Z., ALDAPE, K., COBB, M. H., XIE, K., HE, J., AND LU, Z. FAK phosphorylation by ERK primes ras-induced tyrosine dephosphorylation of FAK mediated by PIN1 and PTP-PEST. *Molecular cell 35*, 1 (July 2009), 11–25.

[172] ZHENG, Y., YANG, W., XIA, Y., HAWKE, D., LIU, D. X., AND LU, Z. Ras-induced and extracellular signal-regulated kinase 1 and 2 phosphorylation-dependent isomerization of protein tyrosine phosphatase (PTP)-PEST by PIN1 promotes FAK dephosphorylation by PTP-PEST. *Molecular and cellular biology 31*, 21 (Nov. 2011), 4258–69.

[173] ZHU, J., ZHU, J., AND SPRINGER, T. A. Complete integrin headpiece opening in eight steps. *The Journal of cell biology 201* (2013), 1053–68.

[174] ZIEGLER, W. H., GINGRAS, A. R., CRITCHLEY, D. R., AND EMSLEY, J. Integrin connections to the cytoskeleton through talin and vinculin. *Biochemical Society transactions 36*, Pt 2 (Apr. 2008), 235–9.

List of Figures

1.1	I-domain of the integrin receptor	6
1.2	Integrin activation	8
1.3	Focal adhesion proteins	11
1.4	Nascent adhesion formation	14
1.5	Maturation process of focal adhesions	16
3.1	Jablonski diagram	24
3.2	Diffraction barrier	26
3.3	Evanescent wave	28
3.4	PALM procedure	30
5.1	Super-resolution imaging set-up	40
6.1	PALM image of mEos2-Zyxin	50
6.2	Degree distribution of simulations	51
6.3	Degree distribution of adhesion proteins	53
6.4	Degree Distribution of wild type and mutated $\beta3$-integrin	57
6.5	Expression level in focal adhesions	58
6.6	Polarity of single focal adhesions	59
6.7	Proximity Plot of Paxillin to compare the radius	61
6.8	Percentage distribution of density	64
6.9	Proximity plot of Kindlin1	69
7.1	Effect of Y27 incubation	75
7.2	Mobility changes of $\beta3$-integrin in focal adhesions	77
7.3	Alterations in the degree distribution during Y27 incubation	79
7.4	Mobile fraction of $\beta3$-integrin during Y27 incubation	80
7.5	Spatial changes in diffusion and density upon Y27 incubation	81
7.6	Density alterations upon Y27 incubation.	82
7.7	Mn^{2+} ions induce clustering and cell spreading.	84
7.8	Kindlin2 response on Mn^{2+} ions	85
7.9	$\beta3$-integrin membrane clustering upon Mn^{2+} treatment	86

7.10 Formation of new adhesion sites 88
7.11 Proximity plots of newly formed adhesion sites 89

8.1 Spreading efficiency of adhesion proteins 92
8.2 Recruitment of FAK and Paxillin to dense domains 94
8.3 Overlay of the adhesion site and recruited Paxillin molecules . 95

List of abbreviations

ADMIDAS adjacent to metal-ion-dependent adhesive site
ARF ADP ribosylation factor
ARP actin related proteins
ASAP1 ARF-GTPase-activating protein 1
cf. *confer*, compare to
CIP calf intestinal alkaline phosphatase
CMV cytomegalovirus
CytD CytochalasinD
D diffusion constant
DBSCAN density-based spatial clustering of applications with noise
DMEM Dulbecco's modified eagle's medium
DMSO dimethyl sulfoxide
ECM extra-cellular matrix
EDTA ethylenediaminetetraacetic acid
e.g. *exempli gratia*, for example
EM-CCD electron multiplying charge-coupled device
FA focal adhesion
F-actin fibrillar actin
FAK focal adhesion kinase
FAT focal adhesion target
FERM-domain Four-point-one, ezrin, radixin, moesin domain
FPALM fluorescence photoactivation localization microscopy
GRAF GTPase regulator associated with FAK
GAP GTPase activating protein
GEF guanine nucleotide exchange factor
GRB Growth factor receptor-bound protein
GTP guanosine triphosphate
HBSS Hanks Balanced Salt solution

HeLa Henrietta Lacks
I-domain inserted domain
LD leucine-aspartate repeat
LIM Lin11, Isl-1 Mec-3 (double-zinc finger motif)
MIDAS metal-ion-dependent adhesive site
MSD mean square displacement
N-WASP neuronal WiskottAldrich syndrome protein
NEB New England Biolabs
PALM photoactivated Localization Microscopy
PA photo-activatable
PBS phosphate buffered saline
PFA paraformaldehyde
PIP2 phosphoinositol 4,5-bisphosphate
PIP3 phosphatidylinositol 3,4,5-triphosphate
PIPKIγ phosphatidylinositol phosphate kinase isoform γ
PSF point spread function
PSI plexin-semaphorin-integrin
PTB phosphotyrosine-binding
REF rat embryo fibroblasts
RGD recognition site of integrin receptors
RIAM Rap1-GTP-interacting adaptor molecule
ROCK Rho-associated protein kinase
ROI region of interest
SH-domain Src homology domain
SPT single particle tracking
STORM stochastic optical reconstruction microscopy
SyMBS synergistic metal ion-binding dependent adhesion site
TIRF total internal reflection fluorescence
TMD transmembrane domain
UV ultraviolet
Y27 Y-27632

i want morebooks!

Buy your books fast and straightforward online - at one of world's fastest growing online book stores! Environmentally sound due to Print-on-Demand technologies.

Buy your books online at
www.get-morebooks.com

Kaufen Sie Ihre Bücher schnell und unkompliziert online – auf einer der am schnellsten wachsenden Buchhandelsplattformen weltweit! Dank Print-On-Demand umwelt- und ressourcenschonend produziert.

Bücher schneller online kaufen
www.morebooks.de

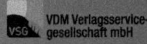 VDM Verlagsservicegesellschaft mbH
Heinrich-Böcking-Str. 6-8   Telefon: +49 681 3720 174   info@vdm-vsg.de
D - 66121 Saarbrücken   Telefax: +49 681 3720 1749   www.vdm-vsg.de

Printed by Books on Demand GmbH, Norderstedt / Germany